Medical
Intelligence
Unit

Immunointervention in Autoimmunity by Th1/Th2 Regulation

Luciano Adorini, M.D.
Roche Milano Ricerche
Milan, Italy

Springer
New York Berlin Heidelberg London Paris
Tokyo Hong Kong Barcelona Budapest

LANDES
BIOSCIENCE
Austin, Texas
U.S.A.

Medical Intelligence Unit

IMMUNOINTERVENTION IN AUTOIMMUNITY
BY TH1/TH2 REGULATION

R.G. LANDES COMPANY
Austin, Texas, U.S.A.

Please address all inquiries to the Publishers:
R.G. Landes Company, 810 S. Church Street, Georgetown, Texas, U.S.A. 78626
Phone: 512/ 863 7762; FAX: 512/ 863 0081

International distributor (except North America):
Springer-Verlag GmbH & Co. KG
Tiergartenstrasse 17, D-69121 Heidelberg, Germany

Springer

International ISBN: 3-540-62733-2

Library of Congress Cataloging-in-Publication Data

Immunointervention in autoimmunity by Th1/Th2 regulation / [edited by] Luciano Adorini.
p. cm. — (Medical intelligence unit)
Includes bibliographical references and index.
ISBN 1-57059-442-2 (alk. paper)
1. Autoimmune diseases--Pathogenesis--Animal models. 2. Th1 cells. 3. Th2 cells. 4. Interleukins--Therapeutic use. 5. Mice--Immunology. I. Adorini, L. (Luciano) II. Series.
[DNLM: 1. Th1 Cells--physiology. 2. Th2 Cells--physiology. 3. Autoimmunity--physiology. 4. Autoimmune Diseases--therapy. QW 568 I3178 1997]
RC600.I44 1997
616.97'8071—dc21
DNLM/DLC
for Library of Congress 95-23860
CIP

Publisher's Note

R.G. Landes Bioscience Publishers produces books in six Intelligence Unit series: *Medical, Molecular Biology, Neuroscience, Tissue Engineering, Biotechnology* and *Environmental.* The authors of our books are acknowledged leaders in their fields. Topics are unique; almost without exception, no similar books exist on these topics.

Our goal is to publish books in important and rapidly changing areas of bioscience for sophisticated researchers and clinicians. To achieve this goal, we have accelerated our publishing program to conform to the fast pace at which information grows in bioscience. Most of our books are published within 90 to 120 days of receipt of the manuscript. We would like to thank our readers for their continuing interest and welcome any comments or suggestions they may have for future books.

Shyamali Ghosh
Publications Director
R.G. Landes Company

CONTENTS

EDITOR

Luciano Adorini, M.D.
Roche Milano Ricerche
Milan, Italy
Preface, Chapter 8

CONTRIBUTORS

Abdallah Badou
Université Paul Sabatier
Hôpital Purpan
Toulouse, France
Chapter 3

Stefan Brocke
National Institute of Neurological Sciences
National Institutes of Health
Bethesda, Maryland, U.S.A.
Chapter 7

Mercy Prabhu Das
Center for Neurologic Disease
Harvard Medical School
Boston, Massachusetts, U.S.A.
Chapter 6

Philippe Druet
Université Paul Sabatier
Hôpital Purpan
Toulouse, France
Chapter 3

Tieno Germann
Institut für Immunologie
Mainz, Germany
Chapter 2

David A. Hafler
Center for Neurologic Disease
Harvard Medical School
Boston, Massachusetts, U.S.A.
Chapter 6

Victoria Heath
Sir William Dunn School of Pathology
University of Oxford
Oxford, U.K.
Chapter 4

Nancy A. Hosken
Immunology Department
DNAX Research Institute
Palo Alto, California, U.S.A.
Chapter 1

Vijay K. Kuchroo
Center for Neurologic Disease
Harvard Medical School
Boston, Massachusetts, U.S.A.
Chapter 6

Vipin Kumar
Department of Microbiology and Molecular Genetics
University of California
Los Angeles, California, U.S.A.
Chapter 5

Don Mason
Sir William Dunn School of Pathology
University of Oxford
Oxford, U.K.
Chapter 4

Lindsay B. Nicholson
Center for Neurologic Disease
Harvard Medical School
Boston, Massachusetts, U.S.A.
Chapter 6

Anne O'Garra
Immunology Department
DNAX Research Institute
Palo Alto, California, U.S.A.
Chapter 1

Lucette Pelletier
Université Paul Sabatier
Hôpital Purpan
Toulouse, France
Chapter 3

Francisco Ramírez
Sir William Dunn School
of Pathology
University of Oxford
Oxford, U.K.
Chapter 4

Abdelhadi Saoudi
Sir William Dunn School
of Pathology
University of Oxford
Oxford, U.K
Chapter 4

Ben Seddon
Sir William Dunn School
of Pathology
University of Oxford
Oxford, U.K.
Chapter 4

Eli Sercarz
Department of Microbiology
and Molecular Genetics
University of California
Los Angeles, California, U.S.A.
Chapter 5

Lawrence Steinman
Department of Immunology
The Weizmann Institute
of Science
Rehovot, Israel
and
Departments of Neurology
and Neurological Sciences
Stanford University
Stanford, California, U.S.A.
Chapter 7

Phil Stumbles
Sir William Dunn School
of Pathology
University of Oxford
Oxford, U.K.
Chapter 4

Ari Waisman
Department of Immunology
The Weizmann Institute
of Science
Rehovot, Israel
and
Departments of Neurology
and Neurological Sciences
Stanford University
Stanford, California, U.S.A.
Chapter 7

Anja Windhagen
Center for Neurologic Disease
Harvard Medical School
Boston, Massachusetts, U.S.A.
Chapter 6

PREFACE

Differentiated T cells produce a restricted set of lymphokines, allowing their subdivision into two major subsets: Th1 and Th2 cells. This has lead to a new paradigm for immunoregulation based on the Th1/Th2 dichotomy. Ten years after the original description, the distinction in type 1 and type 2 T cell responses still holds, and it is currently used as a paradigm to conceptualize pathogenic and protective immune responses. A strict compartmentalization of T cells into Th1 and Th2 is clearly an oversimplification: regulatory and effector mechanisms in the immune system encompass much more than Th1 and Th2 cells. This oversimplification is nevertheless useful for experiments designed to test the paradigm. Based on results obtained in different experimental models of autoimmune diseases, the subdivision of T cells into Th1 and Th2 subsets has been extended to suggest that Th1 cells contribute to the pathogenesis of several organ-specific autoimmune diseases, whereas Th2 cells may inhibit disease development.

A similar dichotomy applies also to human autoimmune diseases. It will soon be possible to formally test immunointervention based on Th1/Th2 cell manipulation in clinical situations: the tools and a conceptual frame are already available. Contributions in this volume will examine key factors affecting the Th1/Th2 balance, in particular antigens and cytokines influencing the development of Th1 and Th2 cells. The rational manipulation of these two variables may ultimately lead to an effective control of Th1 and Th2 cells, potentially able to alter the natural course of autoimmune diseases. Immunointervention strategies based on this paradigm could also be applied to a variety of other pathological conditions, including allergies, infectious diseases and cancer.

CHAPTER 1

ROLE OF CYTOKINES IN THE DEVELOPMENT AND REGULATION OF T HELPER SUBSETS

Anne O'Garra and Nancy A. Hosken

INTRODUCTION

It has been recognized for over 20 years that differential immune responses can be elicited in response to antigen, depending on the form or dose of the antigen, or the route by which it is encountered.[1-3] The discovery that $CD4^+$ T cells could be distinguished according to their ability to produce discrete patterns of cytokines offered a logical explanation for this phenomenon.[4,5] Th1 cells, by their production of IFN-γ and lymphotoxin, could mediate cell-mediated immune responses, whereas the Th2 subset produced cytokines, including IL-4, IL-5, IL-10 and IL-13, are renowned for their role in humoral immunity.[4-7] Th1 responses have been shown to be essential for the eradication of many infectious organisms, but may also result in immunopathology if dysregulated, or result from organ-specific autoimmune disease.[4-13] Since cytokines produced by Th2 cells can activate mast cells and eosinophils, and in addition can result in elevated levels of IgE,[6] they have been strongly implicated in allergy and atopy.[7,14,15] Thus the ability to direct and/or regulate the production of particular cytokines by T helper cells could be of extreme value not only in eliciting the appropriate type of response to a vaccine or invading pathogen, but in addition for intervention during cases of chronic infectious disease, autoimmune pathologies as well as atopy and allergy. A first clue came from observations that cell-mediated and humoral immune responses were often mutually exclusive,[1-3] and a molecular mechanism was suggested by findings showing that the Th2-type cytokines IL-4 and IL-10 could downregulate macrophage activation and bactericidal activity, as well as the development and potentiation of Th1 driven cell-mediated immune responses.[6,16-20]

Immunointervention in Autoimmunity by Th1/Th2 Regulation, edited by Luciano Adorini.

One feature which should be considered with respect to Th1 and Th2 responses is that they result from chronic immunization or chronic disease, and mouse Th1 and Th2 clones were generally isolated from hyperimmunized mice. This explains why such populations were originally difficult to obtain from human systems, since such attempts involved stimulation of peripheral blood T cells from healthy individuals with mitogens.[7,21,22] Indeed, Th1 and Th2 type populations have now been isolated from peripheral blood and draining lymph nodes during chronic infectious diseases and allergy.[7,10,21-23] Helper T cells and clones have been described in human and mouse systems, which produce both Th1 and Th2 type cytokines.[6,7,21,22,24-27] Whether the role of these cells, which have been termed Th0 cells, is as a precursor for Th1 and Th2 cells, or whether Th0 cells represent a separate, stably differentiated population, remains unclear.[28] It is highly likely that Th0 cells may be instrumental in eliminating many pathogens where a balance of regulated cell-mediated immunity, and an appropriate humoral response, will eradicate an invading pathogen with minimum immunopathology. However, chronic conditions may result in polarized Th1 and Th2-type responses, which by the counter-regulatory effects of the cytokines made by the reciprocal subsets, may not only be mutually exclusive but, in addition, pathogenic.[29] To what numerical extent Th1 and Th2 cells dominate such in vivo responses is not yet clear, but their ability to influence chronic disease or pathology through production of high levels of regulatory cytokines is not in doubt (discussed later in the text).

The obvious importance of these CD4$^+$ T cell subsets in immune regulation led a number of researchers, including ourselves, in the last few years, to study the molecular mechanisms leading to the development of such polarized populations. It is clear that the cytokines IL-12 and IL-4 are dominant in driving the development of Th1 and Th2 cells, respectively.[30-36] However, antigen dose and the nature of the antigen presenting cell (APC), have also been suggested to affect the development of these T cell subsets.[37-39] In most cases it is probable that these factors do so by influencing the balance of endogenous levels of IL-12 or IL-4 produced by the APC or the T cell, respectively, thus affecting the subsequent developing Th cell phenotype (Fig. 1.1). Our studies discuss this further and are described below.

ROLE OF CYTOKINES AND ACCESSORY CELLS IN T HELPER SUBSET DEVELOPMENT

Two main types of systems have helped to determine the factors involved in directing Th1 and Th2 phenotype development. The mitogens or antibodies directed against the T-cell receptor (TCR) first used to stimulate 'would-be' naive CD4$^+$ T cells derived from healthy individuals or unimmunized mice, in the presence of accessory signals, or well-defined cytokines, to determine their role in directing subset development.[30,31,36,40,41] The advantage of this is that the role of discrete factors in phenotype development can be studied in the absence of antigen presenting cells (APC), thus verifying that their effect is directly on the T cell. However, the use of mitogens or anti-TCR antibodies to stimulate $CD4^+$ T cells in the periphery may induce responses of mixtures of memory antigen-specific T cells as well as naive T cells. More recently, in an attempt to represent a more physiological antigen-specific system, we and others have used an antigen-specific αβ-TCR transgenic mouse as a source of naive $CD4^+$ T cells, with a defined antigen specificity, to study Th phenotype development.[32-35,42,43] Although the presence of APC in the system inevitably complicates the read-out, this facilitates the identification of other cofactors which may affect subset development.

IL-12 INDUCED TH1 DEVELOPMENT

IL-12 is a dominant factor in directing the development of Th1 cells producing high levels of IFN-γ, required for eliciting a cell-mediated immune response responsible for the eradication of numerous pathogens.[33,35,36,44-47] IL-12 is a 75 kDa heterodimer shown to be secreted by monocytes[47-50] with multiple activities on natural killer (NK) cells and T cells,[47] including augmentation of IFN-γ production.[47,51] Although mice in which the IL-12 p40 gene has been targeted show reduced levels of IFN-γ and DTH responses, these responses are not completely absent, suggesting that there are alternative mechanisms for the development of Th1 cells producing IFN-γ.[52] However, that this factor plays a dominant role in the development of IFN-γ producing cells capable of eliciting a response sufficient for the clearance of pathogens such as *Leishmania major*, is supported by susceptibilty to this pathogen in dual IL-12 p35 and p40 gene-targeted mice,[53] and by the ability of IL-12 to render susceptible mice resistant to this pathogen.[54-56]

CELLULAR SOURCES OF IL-12 AND THEIR ROLE IN TH1 PHENOTYPE DEVELOPMENT

Antigen presenting cells, such as macrophages, had been originally shown to direct Th1 development by secretion of IL-12.[33,57] We showed that dendritic cells induced strong antigen-specific proliferation of $CD4^+$ T cells from mice expressing an OVA-specific transgenic TCR, but addition of *Listeria*-activated macrophages producing high levels of IL-12, or large amounts of exogenous IL-12, was required for Th1 development.[33,57] The added combination of IL-4 and IL-12 during primary activation of the $CD4^+$ T cells led to Th2 development, suggesting that the effect of IL-4 was dominant over IL-12 in the development of Th phenotypes.[33] Moreover, IL-10's ability to inhibit the development of Th1 cells, as well as the induction of IFN-γ by Th1 clones, can now be

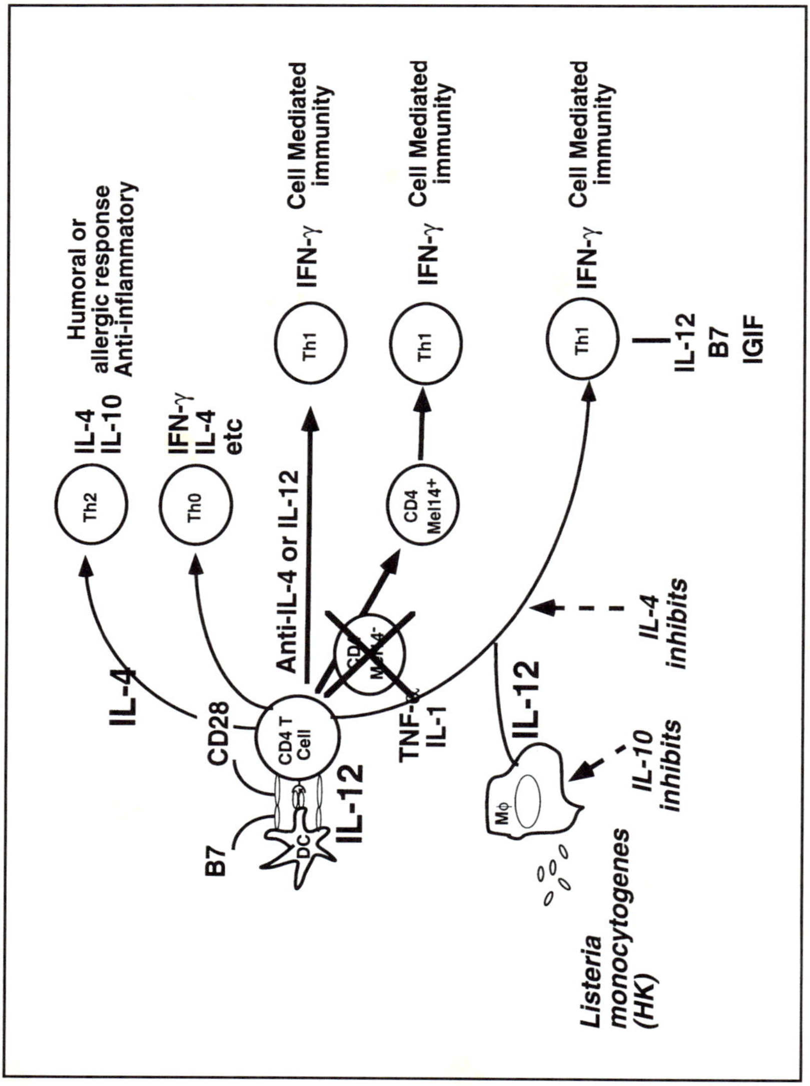

Fig. 1.1. Model for cytokine-driven Th1 and Th2 development. Reprinted with permission from O'Garra A, Murphy K. Chem Immunol. ©Karger.

attributed to its inhibition of IL-12 production by macrophages[58-60] as well as down-regulation of membrane-bound costimulators[59,61] and Class II MHC antigens.[62]

Our original studies[57] suggested that dendritic cells, although capable of expressing high levels of membrane-bound accessory or costimulatory molecules, did not produce cytokines necessary for induction of discrete Th subsets from naive T cells. However, upon further investigation, it became clear that dendritic cells can drive the development of IFN-γ producing Th1 cells from $CD4^+$ T cells primed in the absence of IL-4.[63] Primarily, dendritic cells induced the development of Th1 cells producing high levels of IFN-γ and undetectable IL-4 from total FACS-purified, OVA-specific $CD4^+$ T cells upon neutralization of endogenous levels of IL-4. This Th1 development resulting from the addition of anti-IL-4 mAb was significantly inhibited by anti-IL-12 mAbs added during primary T cell activation with Ag and dendritic cells as APC. A possible source of this endogenous IL-4 was the LECAM-1^{dull} subset of $CD4^+$ T cells, which contains the majority of the "memory/activated $CD4^+$ T cells and has been shown to secrete significant amounts of IL-4 upon primary stimulation in vitro, in contrast to the "naive" LECAM-1^{bright} subset.[64,65] Indeed, upon removal of this LECAM-1^{dull}, $CD4^+$ population which produced detectable levels of IL-4 upon primary stimulation, dendritic cells were then able to drive IL-12-dependent development of Th1 cells in the presence of Ag alone.[63] Intracellular staining for the p40 subunit of IL-12 with FITC-labelled anti-p40 mAbs, confirmed that dendritic cells coexpressing the NLDC-145 antigen produced the IL-12 protein, and this appeared to be enhanced by previous incubation with TCR-transgenic CD4+ T cells plus their specific antigen, OVA.[63]

These data suggested that dendritic cells may produce levels of IL-12 insufficient to counteract the inhibition of Th1 development by endogenous IL-4 produced by the LECAM-1^{dull}, $CD4^+$ population. Thus, the relative levels of T cell derived IL-4 versus APC (macrophage or dendritic cell)-derived IL-12 strictly control whether Th1 cells responsible for cell-mediated immunity develop in response to antigen. This may be determined by the nature of the antigenic stimulus, which may or may not be able to induce IL-12 from the APC, as well as the dose of antigen which may determine the levels of IL-4 induced by the resulting T cells.[66]

In contrast to dendritic cells, B cells, even when significantly activated so that they can induce similar levels of proliferation as dendritic cells, still drive a Th0 response, with relatively low levels of IFN-γ and IL-4 being produced even upon neutralization of IL-4 or removal of "memory" or "regulatory" LECAM-1^{dull}, $CD4^+$ T cells (ref. 66 and Hosken and O'Garra, unpublished). The low levels of IFN-γ produced by these Th cells developing from $CD4^+$ T cells stimulated with antigen and B cell as APC in primary cultures, did not appear to be IL-12 dependent, suggesting that normal mouse B cells do not produce this cytokine despite the original detection of IL-12 in human B cell lymphoblastoid lines.[44,50,67]

COFACTORS INVOLVED IN IL-12-INDUCED TH1 DEVELOPMENT

In the system described above, we have shown that IL-12-driven Th1 development is dependent on another cytokine, IFN-γ.[33,57,68] The requirement for this factor in the development of Th1 cells has been controversial.[34,35,41,42,57,68-76] This may be explained in part by different genetic backgrounds of mice under study,[77] or our further observations that separation of the $CD4^+$ T cells on the basis of expression of LECAM-1 before stimulation, shows that development of Th1 cells producing IFN-γ from the $CD4^+$, LECAM-1^+, so-called "naive" cells is dependent on IFN-γ, whereas their development from "memory"[64] or "regulatory"[65,78] $CD4^+$, LECAM-1^- is not.[75] Although IFN-γ has been shown to be required for Th1 development in an APC-independent system,[72,79] it is probable that there is an effect via the APC in complex in vivo systems,[76] as well as a direct effect on the T cell. Furthermore, it is possible that IFN-γ's effects on the T cell are to inhibit IL-4 production or action, thus counteracting the inhibitory effect of IL-4 on IL-12-induced Th1 development.

In addition to the dominant effect of IL-12 in driving Th1 development, we have shown that IL-1 and TNF-α are required for IL-12 induction of Th1 cells producing maximal levels of IFN-γ (Macatonia and O'Garra unpublished; Shibuya and O'Garra, in preparation). More recently, we have observed that whereas IL-1 and TNF-α are required for IL-12 responsiveness and subsequent Th1 development from naive $CD4^+$ T cells from the DO11.10, OVA-TCR transgenic mice (BALB/c genetic background), these factors were not required for Th1 development from naive $CD4^+$ T cells from the TCR-7, HEL-specific-TCR transgenic mouse (C57BL/6 genetic background) (Shibuya and O'Garra, in preparation; Hartley and O'Garra, in preparation). To confirm these results in an independent fashion, we measured the induction of CD25, the IL-2 receptor α subunit (IL-2Rα), since it has been shown that IL-12 augments IL-2 induction of IL-2Rα in Th1 clones.[80] This defect in IL-12 responsiveness in the BALB/c OVA-TCR transgenic T cells was also observed independently at the level of induction of the IL-2Rα as previously reported by Güler et al.[81] We now show that IL-12 responsiveness in the OVA-specific TCR transgenic T cells (BALB/c), with respect to IL-2Rα, can also be restored upon addition of IL-1 and TNF-α, whereas expression of the IL-2Rα in the HEL-specific TCR transgenic T cells (C57BL/6) is maximally induced by IL-12 in the absence of these factors (Shibuya and O'Garra, in preparation). Whether the effect of IL-1 and TNF-α is to directly upregulate the IL-12 receptor[82-84] or to upregulate IFN-γ, which may then upregulate the IL-12 receptor, is not clear at this stage. Furthermore, the requirement for these factors in the OVA-specific TCR transgenic mouse (BALB/c) and not the HEL-specific TCR transgenic mouse (C57BL/6) may be related to genetic defects in Th1 development reported for mice on a BALB/c genetic background.[8,9,77,81,85,86]

It has been demonstrated that TGF-β may play a role in the development of Th1 cells,[68,87-90] but thus far its mechanism of action remains unclear, particularly since this factor may also inhibit an ongoing effector Th1-type cell-mediated immune response (refs. 91-94 and Murphy and O'Garra, unpublished).

ENHANCED IFN-γ PRODUCTION FROM COMMITTED TH1 POPULATIONS AND TH1 CLONES

As well as directing the development of Th1 cells from naive CD4+ T cells, IL-12 can also potentiate the production of IFN-γ from already differentiated Th1 populations and Th1 clones.[60] Recently, it has been demonstrated that a recently cloned cytokine in both mouse and human, interferon-gamma inducing factor (IGIF), can also enhance the production of IFN-γ by anti-CD3-stimulated cloned murine Th1 cells[95] and enriched human peripheral blood T cells.[96] This factor has been postulated to be responsible for the residual Th1 development observed in IL-12 p40 gene-targeted mice;[52] however, in our hands this factor does not drive Th1 development, but only potentiates the production of IFN-γ from differentiated Th1 cells and clones, particularly in synergy with IL-12 (Shibuya and O'Garra, in preparation). IGIF has been suggested to belong to the IL-1 family of cytokines based on its structural similarity to IL-1,[97] but in contrast to IL-1α, IGIF (IL-1γ) does not costimulate for IL-12-induced Th1 development (Shibuya and O'Garra, in preparation). Messenger RNA encoding IGIF has been detected in activated Kupffer cells and macrophages,[95] but what other cells may produce this factor and how it may be regulated remain unclear.

IL-4 DIRECTS TH2 DEVELOPMENT

Using this reductionist approach it became clear that the addition of IL-4 during primary culture of antigen or mitogen-stimulated CD4+ T cells resulted in the development of Th2 cells producing high levels of IL-4 and undetectable levels of IFN-γ upon subsequent restimulation.[32,34] IL-4-induced Th2 development was independent of the APC used to stimulate the T cells with antigen, and thus far has not been achieved with any other accessory signal/cytokine.[32,34]

The requirement for IL-4 during priming has also been shown to play a major role in vivo in directing a Th2 response to microbial pathogens including *Leishmania major* and *Candida albicans*.[98-101] Furthermore, nematode infection of mice lacking a functional IL-4 gene, as a result of functional inactivation of the gene by homologous recombination, resulted in much reduced levels of the Th2-type cytokines IL-5, IL-9 and IL-10 from $CD4^+$ T cells than the wild-type control mice.[101] However, although there was a reduction in the levels of Th2 cytokines, as well as the eosinophilia characteristic of a Th2-type response, in keeping with IL-4's dominant role, this response was not completely absent.[101] Surprisingly, in view of earlier studies performed with anti-IL-4 antibodies (see above)[98] BALB/c mice, whose IL-4 gene has been inactivated by gene targeting, are still unable to eradicate the parasite *Leishmania major*,[102] although at present there is much controversy surrounding this issue.

ROLE OF ANTIGEN DOSE IN DIRECTING T HELPER PHENOTYPE DEVELOPMENT

It has been recently demonstrated that differentiated effector $CD4^+$ T cells may produce different cytokines depending on the dose of antigen used.[37,39] Until very recently, however, there had been no clear study demonstrating that the dose of antigen can directly induce the development of Th1 or Th2 phenotypes from naive $CD4^+$ T cells, although this has been suggested.[38]

We have now used naive $CD4^+$ T cells from the DO11.10 TCR-transgenic mouse as described above to determine whether different doses of peptide antigen will preferentially stimulate the development of Th1 or Th2 cells. We demonstrate that at doses of antigen that will induce seemingly similar levels of T cell proliferation, low to medium doses of antigen will result in Th1 cells producing IFN-γ and undetectable levels of IL-4, whereas increasing the dose of antigen will result in the disappearance of IFN-γ and the development of IL-4 producing cells.[66] Coincidental with this disappearance of IFN-γ producing cells is the observation of massive cell death in the cultures. This trend of differential IFN-γ versus IL-4 producing Th cell development is still observed even when antigen dose is varied in primary cultures of high density $CD4^+$, LECAM-1^+ T cells. This suggests that it is not due to a small contaminating population of IL-4 producing cells which will be evident at higher doses of antigen, although this cannot be completely ruled out. At extremely low doses of antigen, when there is a poor recovery of T cells after culture of high density $CD4^+$, LECAM-1^+ T cells, IL-4-production is dominant over IFN-γ, again suggesting possible contamination with a subset of $CD4^+$ T cells with enhanced viability. In our system, the development of Th2 cells resulting from either high or low doses of antigen was completely abrogated by the addition of neutralizing anti-IL-4 antibodies. This effect of low dose of antigen on preferential Th2 development is in keeping with studies published in parallel by Constant et al[103] who, however, did not obtain Th2 development at the highest doses of antigen shown in their study. This could be due to differences in the antigen used and perhaps not achieving a high enough dose in their system, or as a result of the genetic background differences. With respect to genetic background and type of antigen, we have recently shown that Th2 development resulting from both high and low dose of antigen, as well as Th1 development in response to median doses of antigen, can also be observed in mice on a C57BL/6 background with a transgenic-TCR specific for HEL (Hosken and O'Garra, in preparation). This effect of antigen dose in directly inducing the development of Th1 or Th2-type T cells could be observed regardless of whether dendritic cells or B cells were used as APC. However, it is likely than in normal antigen-specific T cells, where a much lower frequency of cells recognize the same antigen that there will be a very fine balance between APC producing cytokines such as IL-12, and $CD4^+$ T cells producing IL-4, and that the dose of antigen will be critical in determining the development of IFN-γ versus IL-4-producing cells. Undoubtedly, in vivo, the form of antigen as well as the dose will have an important role in determining Th phenotype development, particularly since particulate antigens, and micro-organisms, will activate macrophages to produce high levels of IL-12 and favor the development of Th1 responses.[33,57,63,68] The mechanisms underlying antigen dose driven Th phenotype development remain unclear. However, it is of note that antigen dose driven Th1 and Th2 phenotype development could be completely overridden by the addition of saturating amounts of IL-4 or IL-12 respectively, to the primary cultures (Hosken and O'Garra, in preparation).

These studies, on the role of antigen dose affecting Th cell phenotype development, raise concerns regarding the use of antigenic proteins or peptides, as opposed to attenuated organisms as vaccines, since it will be impossible to predict an appropriate dose for immunization of a genetically diverse human population. The ability of cytokines, and perhaps adjuvants, to help achieve the appropriate response to a protein/peptide antigen may help to overcome this problem, and lead to enhanced protection.

STABILITY AND MAINTENANCE OF T HELPER CELL PHENOTYPES

Can the Phenotype of a Th1 Population of $CD4^+$ T Cells Be Changed?

Since a large of number of organ-specific autoimmune diseases have been associated with an inflammatory Th1-response, the ability to convert a Th1 response to a less inflammatory Th2 or regulated response, would be of great therapeutic benefit.[11,12,104-106] That Th1 populations can give rise to Th2-like cells has been suggested from observations in human systems which demonstrated that purified protein derivative (PPD)-specific T cells which normally outgrow with a Th1 like profile, can be selected as Th2-like if IL-4 is added early in bulk cultures.[71] In addition, polarized Th1 populations obtained from mice infected with *L. major* have been modified to produce Th2-type cytokines by reculture in vitro with specific antigen, APC, IL-2 and IL-4.[107] Furthermore, populations of Th1 cells derived from naive TCR-transgenic $CD4^+$ T cells by one round of stimulation with antigen and APC plus IL-12 have also been converted to Th2 populations by one round of secondary stimulation with antigen and APC plus exogenous IL-4.[84,108] Although even in the latter cases where T cells were clonal with respect to their antigen receptor, these systems undoubtedly all deal with heterogeneous populations of Th cells with respect to their cytokine production,[27,109] and the level of commitment for cytokine production at the single cell level is unclear. Polarized Th1 populations derived from naive TCR-transgenic $CD4^+$ T cells after one round of antigenic stimulation in the presence of IL-12, appeared homogeneous as assessed by immunoassay.[109] However, we have shown by analysis for cytokine production at the single cell level using flow cytometry that these populations are indeed heterogeneous.[109] Thus the ability to convert a Th1 to a Th2 population may result from Th2 development from uncommitted precursor T cells or expansion of contaminating, previously committed Th2 cells. With respect to conversion of a Th1 to a Th2 population this is greatly facilitated by the disappearance of IFN-γ producing cells after one round of stimulation with antigen plus APC in the presence of IL-4.[109] Whether this is a direct effect on the T cell or occurs via the APC is not clear. We further support that this Th1 to Th2 switch is as a result of heterogenaeity at the population level, since Th1 clones, or more homogeneous Th1 populations derived from chronic stimulation with antigen under Th1-polarizing conditions, could not be converted to the opposite phenotype.[109] Chronic stimulation of these seemingly committed Th1 populations with antigen and APC in the presence of IL-4 could significantly reduce the levels of IFN-γ produced by such Th1 cells when restimulated with antigen in the absence of IL-4, although not giving rise to any IL-4 producing cells.[109]

Can the Phenotype of a Th2 Population of $CD4^+$ T Cells Be Changed?

Th2 responses have been frequently correlated with an inability to eradicate a pathogen, in contrast to Th1 responses which usually correlate with protection, particularly against intracellular pathogens.[8,9,98,99,110] Th2 responses have also been implicated in the pathology associated with allergy and atopy.[7,14] Thus, if it were possible to convert an established Th2 response to a Th1 response, and perhaps obtain homeostasis, this could offer strategies for therapeutic intervention.

Recently, ongoing Th2 responses to *L. major* were found to be reverted to a healing Th1 response when BALB/c mice were treated during chronic infection with Pentostam (to reduce the parasite infection) together with IL-12,[111] or by transferring the Th2 populations to scid mice under appropriate conditions.[112,113]

However, in contrast to the in vivo studies discussed above showing the reversal of a Th2 phenotype, it has been recently shown that, although the phenotype of early Th1 cells was efficiently reversed in vitro at the population level by addition of IL-4 during antigenic stimulation, Th2 cells were resistant to phenotype reversal.[84,108] The inability to convert such Th2 populations derived from one round of antigenic stimulation of TCR-transgenic T cells in IL-4, to Th1 cells by restimulating with antigen in the presence of IL-12, has been suggested to reflect a loss of responsiveness of Th2 cells to IL-12.[84] In agreement with this, in our system, although these short term populations of Th2 cells can give rise to Th1 cells upon antigenic restimulation in IL-12, this is very likely as a result of development from uncommitted precursor T cells or expansion of contaminating, previously committed Th1 cells, since upon analysis at the single cell level the IL-4 producing population remained numerically identical.[109] This is supported by our observations that chronically stimulated Th2 populations and clones cannot give rise to IFN-γ producing cells.[109] These data are in contrast to studies in human systems where short-term Th2 cells derived from cord blood cells upon stimulation with mitogen can still be induced to produce IFN-γ if restimulated in the presence of IL-12, and in addition reduce significantly in number in the IL-12 stimulated cultures.[40] These data are in accordance with the ability of IL-12 to induce IFN-γ production in seemingly Th2 but more likely Th0 clones in the human.[114] This may reflect differences in the level of commitment of the populations in mouse and human systems and/or perhaps expression of the IL-12 receptor. However, these studies in all may suggest that it may be more difficult to change or reduce an IL-4 producing Th2 population than a IFN-γ-producing Th1 population.

IMPLICATIONS FOR IMMUNOTHERAPY IN CHRONIC DISEASES ARISING FROM INFECTION, ALLERGY OR T CELL MEDIATED AUTOIMMUNITY

These studies may provide new approaches for therapeutic intervention during chronic disease. It is likely that even during chronic disease, manifested by high levels of Th1-type or Th2-type cytokines, that the population of $CD4^+$ T cells in a draining lymph node consists of a spectrum of Th cells at different stages of differentiation (Fig. 1.2). Immune regulation in vivo will most likely restrict the frequency of completely differentiated and polarized Th1 and Th2 clones within the population. However, these cells may play an important role in determining the extent of pathology, by their production of high levels of cytokines. Our data suggest that it may be possible to override the controlling influence of such clones producing extreme levels of cytokines. For example, the administration of antigen together with the polarizing conditions will enhance a change in Th phenotype by the recruitment of naive and/or

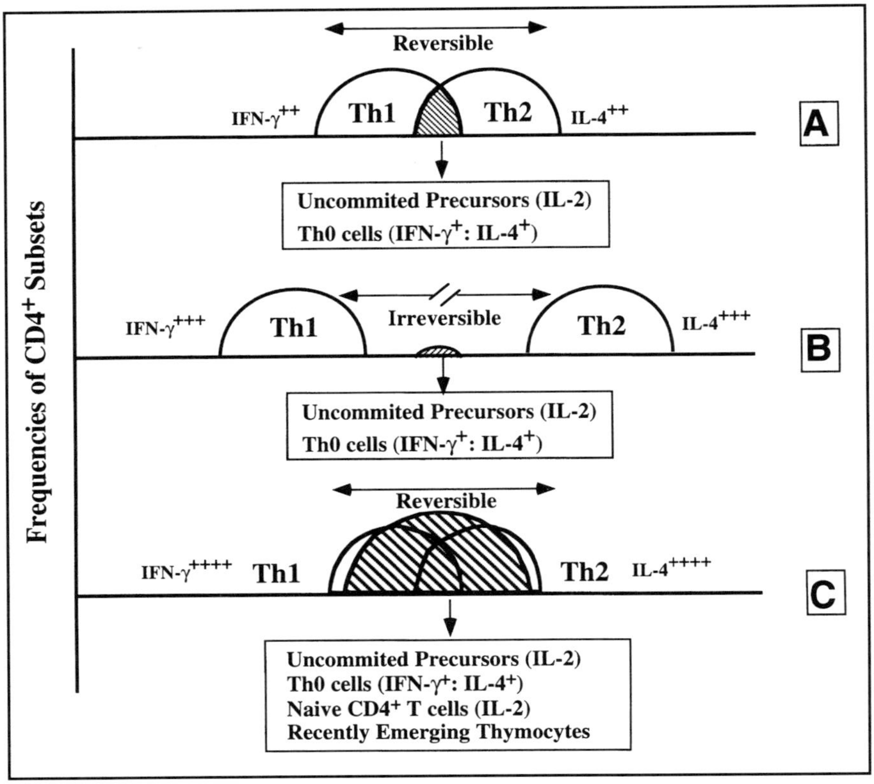

Fig. 1.2. Model for reversal of T helper phenotypes. The model for heterogeneity of Th phenotypes and their reversal is based: (A and B) on data obtained from flow cytometric analysis of intracellular cytokine synthesis in polarized Th1 and Th2 populations obtained from: αβ-TCR transgenic $CD4^+$ T cells stimulated in vitro with APC plus OVA in the presence of IL-12 plus anti-IL-4 mAbs (Th1), or IL-4 (Th2); for (A) 1 week; (B) 3 weeks; (C) the prediction of the heterogeneity of Th phenotypes and their reversal during an immune response in vivo. Reprinted with permission from Murphy et al. J Exp Med 1996. ©Rockefeller University Press.

uncommitted cells, which will undoubtedly be present even during chronic disease. These cells, by differentiation and clonal outgrowth, will suppress the dominant Th phenotype as modeled from our flow cytometric analysis of IFN-γ and IL-4 producing cells in the in vitro reversal of polarized Th cell populations. The ability of IL-4 to downregulate IFN-γ producing cells may indicate that it may be more possible to dampen or revert inflammatory or Th1-mediated pathologies than atopic or chronic Th2-dominated diseases, since IL-4 producing cells were more resistant to change. In these situations it may be necessary to reduce the ongoing immune response as well as to restimulate the populations in the presence of IL-12 plus anti-IL-4 in order to outnumber and suppress dominant clones of a Th2 phenotype.

REFERENCES

1. Parish CR. The relationship between humoral and cell-mediated immunity. Transplant Rev 1972; 13:35-66.
2. Katsura Y. Cell-mediated and humoral immune responses in mice. III. Dynamic balance between delayed-type hypersensitivity and antibody response. Immunology 1977; 32:227-235.
3. Asherson GL, Stone SH. Selective and specific inhibition of 24-hour skin reactions in the guinea pig. I. Immune deviation: description of the phenomenon and the effect of splenectomy. Immunology 1965; 9:205-217.
4. Mosmann TR, Cherwinski H, Bond MW, Giedlin MA, Coffman RL. Two types of murine helper T cell clone. I. Definition according to profiles of lymphokine activities and secreted proteins. J Immunol 1986; 136:2348-2357.
5. Mosmann TR, Coffman RL. Th1 and Th2 cells: different patterns of lymphokine secretion lead to different functional properties. Ann Rev Immunol 1989; 7:145-173.
6. Sher A, Coffman RL. Regulation of immunity to parasites by T cells and T cell-derived cytokines. Ann Rev Immunol 1992; 10:385-409.
7. Romagnani S. Human T_H1 and T_H2 subsets: doubt no more. Immunol Today 1991; 12:256-257.
8. Scott P, Natovitz P, Coffman RL, Pearce E, Sher A. Immunoregulation of cutaneous leishmaniasis. T cell lines that transfer protective immunity or exacerbation belong to different T helper subsets and respond to distinct parasite antigens. J Exp Med 1988; 168:1675-1684.
9. Heinzel FP, Sadick MD, Holaday BJ, Coffman RL, Locksley RM. Reciprocal expression of interferon gamma or interleukin 4 during the resolution or progression of murine leishmaniasis. J Exp Med 1989; 169:59-72.
10. Salgame P, Abrams JS, Clayberger C, Goldstein H, Convit J, Modlin RJ, Bloom BR. Differing lymphokine profiles of functional subsets of human CD4 and CD8 T cell clones. Science 1991; 254:279-282.
11. O'Garra A, Murphy K. T-cell subsets in autoimmunity. Curr Opin Immunol 1993; 5:880-886.
12. Powrie F, Coffman RL. Cytokine regulation of T-cell function: potential for therapeutic intervention. Immunol Today 1993; 14:270-274.
13. Fowell D, Mason D. Evidence that the T cell repertoire of normal rats contains cells with the potential to cause diabetes. Characterization of the CD4+ T cell subset that inhibits this autoimmune potential. J Exp Med 1993; 177:627-636.
14. Robinson DS, Hamid Q, Ying S, Tsicopoulos A, Barkans J, Bentley AM, Corrigan CJ, Durham SR, Kay AB. Evidence for a predominant "Th2-

type" bronchoalveolar lavage T-lymphocyte population in atopic asthma. New Engl J Med 1992; 326:298-304.
15. Romagnani S. Regulatory role of IL-4 and other cytokines on the function and development of human T cell clones. Res Immunol Immunobiol of IL-4. 1993; 144:625-628.
16. Fiorentino DF, Zlotnik A, Vieira P, Mosmann TR, Howard M, Moore KW, O'Garra A. IL-10 acts on the antigen-presenting cell to inhibit cytokine production by Th1 cells. J Immunol 1991; 146:3444-3451.
17. Fiorentino DF, Zlotnik A, Mosmann TR, Howard M, O'Garra A. IL-10 inhibits cytokine production by activated macrophages. J Immunol 1991; 147:3815-3822.
18. de Waal Malefyt R, Abrams J, Bennett B, Figdor C, Vries de JE. IL-10 inhibits cytokine synthesis by human monocytes: An autoregulatory role of IL-10 produced by monocytes. J Exp Med 1991; 174:1209-1220.
19. Powrie F, Coffman R. IL-4 and IL-10 inhibit DTH and IFN-γ production. Eur J Immunol 1993; 23:2223-2229.
20. Gazzinelli RT, Oswald IP, James SL, Sher A. IL-10 inhibits parasite killing and nitric oxide production by IFN-γ activated macrophages. J Immunol 1992; 148:1792-1796.
21. Andersson U, Andersson J, Lindfors A, Wagner K, Möller G, Hewsser CH. Simultaneous production of interleukin 2, interleukin 4 and interferon-γ by activated human blood lymphocytes. Eur J Immunol 1990; 20:1591-1596.
22. de Vries JE, de Waal Malefyt R, Yssel H, Roncarolo M-G, Spits H. Do human TH1 and TH2 $CD4^+$ clones exist? $CD4^+$ T-cell Subsets: Differentiation and Function 1991; 142:59-63.
23. Modlin RL, Gersuk GM, Nelson EE, Pattengale PK, Gunter JR, Chen L, Cooper CL, Bloom BR, Rea TH. T-lymphocyte clones from leprosy skin lesions. Lepr Rev 57, suppl 1986; 2:143.
24. Mosmann TR, Coffman RL. Heterogeneity of cytokine secretion patterns and functions of helper T cells. 1989; Adv Immunol 46:111-147.
25. Kelso A, Gough NM. Coexpression of granulocyte-macrophage colony-stimulating factor, γ interferon, and interleukins 3 and 4 is random in murine alloreactive T-lymphocyte clones. Proc Natl Acad Sci USA 1988; 85:9189-9193.
26. Firestein GS, Roeder WD, Laxer JA, Townsend KS, Weaver CT, Hom JT, Linton J, Torbett BE, Glasebrook AL. A new murine CD4+ cell subset with an unrestricted cytokine profile. J Immunol 1989; 143:518-525.
27. Openshaw P, Murphy E, Hosken NA, Maino V, Davis K, Murphy K, O'Garra A. Heterogeneity of intracellular cytokine synthesis at the single cell level in polarized T helper 1 and T helper 2 populations. J Exp Med 1995; 182:1357-1367.
28. Kamogawa Y, Minasi L-AE, Carding SR, Bottomly K, Flavell RA. The relationship of IL-4 and IFN-γ-producing T cells studied by lineage ablation of IL-4 producing cells. Cell 1993; 75:985-995.

29. Alwan WH, Kozlowska WJ, Openshaw PJM. Distinct types of lung disease caused by functional subsets of antiviral T cells. J Exp Med 1994; 179:81-89.
30. Swain SL, Weinberg AD, English M, Huston G. IL-4 directs the development of Th2-like helper effectors. J Immunol 1990; 145:3796-3806.
31. Le Gros G, Ben-Sasson S, Seder R, Finkelman FD, Paul WE. Generation of interleukin 4 (IL-4)-producing cells in vivo and in vitro: IL-2 and IL-4 are required for in vitro generation of IL-4 producing cells. J Exp Med 1990; 172:921-929.
32. Hsieh C-S, Heimberger AB, Gold JS, O'Garra A, Murphy K. Differential regulation of T helper phenotype development by IL-4 and IL-10 in an αβ-transgenic system. Proc Natl Acad Sci USA 1992; 89:6065-6069.
33. Hsieh C-S, Macatonia SE, Tripp CS, Wolf SF, O'Garra A, Murphy KM. Development of Th1 CD4+ T cells through IL-12 produced by Listeria-induced macrophages. Science 1993; 260:547-549.
34. Seder RA, Paul WE, Davis MM, Fazekas de St. Groth B. The presence of interleukin 4 during in vitro priming determines the lymphokine-producing potential of $CD4^+$ T cells from T cell teceptor transgenic mice. J Exp Med 1992; 176:1091-1098.
35. Seder RA, Gazzinelli R, Sher A, Paul WE. Interleukin 12 acts directly on $CD4^+$ T cells to enhance priming for interferon γ production and diminishes interleukin 4 inhibition of such priming. Proc Natl Acad Sci USA 1993; 90:10188-10192.
36. Manetti R, Parronchi P, Guidizi MG, Piccinni M-P, Maggi E, Trinchieri G, Romagnani S. Natural killer cell stimulatory factor (interleukin 12 [IL-12]) induces T helper type 1 (Th1)-specific immune responses and inhibits the development of IL-4-producing cells. J Exp Med 1993; 177:1199-1204.
37. Carballido JM, Carballido-Perrig N, Terres G, Heusser CH, Blaser K. Bee venom phospholipase A2 specific T cell clones from human allergic and non-allergic individuals: cytokine patterns change in response to the antigen concentration. Eur J Immunol 1992; 22:1357.
38. Pfeiffer C, Murray J, Madri J, Bottomly K. Selective activation of Th1- and Th2-like cells in vivo—response to human collagen IV. Immunol Rev 1991; 123:65-84.
39. Secrist H, DeKruyff RH, Umetsu DT. Interleukin 4 production by $CD4^+$ T cells from allergic individuals is modulated by antigen concentration and antigen presenting cell type. J Exp Med 1995; 181:1081-1089.
40. Sornasse T, Larenas PV, Davis KA, de Vries JE, Yssel H. Differentiation and stability of Th1 and Th2 cells derived from naive human neonatal $CD4^+$ T cells, analyzed at the single-cell level. J Exp Med 1996; in press.:
41. Maggi E, Parronchi P, Manetti R, Simonelli C, Piccinni M-P, Rugiu FS, De Carli M, Ricci M, Romagnani S. Reciprocal regulatory effects of IFN-gamma and IL-4 on the in vitro development of human Th1 and Th2 clones. J Immunol 1992; 148:2142.

42. Swain SL, Bradley LM, Croft M, Tonkonogy S, Atkins G, Weinberg AD, Duncan DD, Hedrick SM, Dutton RW, Huston G. Helper T-cell subsets: phenotype, function and the role of lymphokines in regulating their development. Immunol Rev 1991; 23:115-144.
43. Swain SL. IL-4 dictates T-cell differentiation. 53rd Forum Immunol Res Immunol 1993; 144:567-651.
44. Kobayashi M, Fitz L, Ryan M, Hewick RM, Clark SC, Chan S, Loudon R, Sherman F, Perussia B, Trinchieri G. Identification and purification of natural killer cell stimulatory factor (NKSF), a cytokine with multiple biologic effects on human lymphocytes. J Exp Med 1989; 170:827-845.
45. Tripp CS, Wolf SE, Unanue ER. Interleukin-12 and tumor necrosis factor α are costimulators of interferon-γ production by natural killer cells in severe combined immunodeficiency mice with listeriosis, and interleukin 10 is a physiologic anatagonist. Proc Natl Acad USA 1993; 90:3725-3729.
46. Gazzinelli RT, Hieny S, Wynn TA, Wolf S, Sher A. Interleukin 12 is required for the T-lymphocyte-independent induction of interferon γ by an intracellular parasite and induces resistance in T-cell deficient hosts. Proc Natl Acad Sci USA 1993; 90:6115-6119.
47. Trinchieri G. Interleukin-12 and its role in the generation of Th1 cells. Immunol Today 1993; 14:335.
48. Gubler U, Chua AO, Schoenhaut DS, Dwyer CM, McComas W, Motyka R, Nabavi N, Wolitzky AG, Quinn PM, Familetti PC, Gately MK. Coexpression of two distinct genes is required to generate secreted bioactive cytotoxic lymphocyte amturation factor. Proc Natl Acad Sci USA 1991; 88:4143-4147.
49. Schoenhaut DS, Chua AO, Wolitzky AG, Quinn PM, Dwyer CM, McComas WP, Familletti C, Gately MK, Gubler U. Cloning and expression of IL-12. Journal of Immunology 1992; 148:3433-3440.
50. D'Andrea A, Rengaraju M, Valiante NM, Chemini J, Kubin M, Aste M, Chan SH, Kobayashi M, Young D, Nickbarg E, Chizzonite R, Wolf SF, Trinchieri G. Production of natural killer cell stimulatory factor (Interleukin 12) by peripheral blood mononuclear cells. J Exp Med 1992; 176:1387-1398.
51. Chan SH, Kobayashi M, Santoli D, Perussia B, Trinchieri G. Mechanisms of IFN-γ induction by natural killer cell stimulatory factor (NKSF/IL-12): Role of transcription and mRNA stability in the synergistic interaction between NKSF and IL-2. J Immunol 1992; 148:92.
52. Magram J, Connaughton SE, Warrier RR, Carvajal DM, Wu C, Ferrante J, Stewart C, Sarmiento U, Faherty DA, Gately MK. IL-12-deficient mice are defective in IFN-γ production and type 1 cytokine responses. Immunity 1996; 4:471-481.
53. Mattner F, Magram J, Ferrante J, Launois P, Di Padova K, Behin R, Gately M, Louis Alber JG. Genetically resistant mice lacking interleukin-

12 are susceptible to infection with Leishmania major and mount a polarized Th2 cell response. Eur J Immunol 1996; 26:1553-1559.
54. Sypek JP, Chung CL, Mayor SEH, Subramanyam JM, Goldman SJ, Sieburth DS, Wolf SE, Schaub RG. Resolution of cutaneous Leishmaniasis: interleukin 12 initiates a protective T helper type 1 immune response. J Exp Med 1993; 177:1797-1802.
55. Heinzel FP, Schoenhaut DS, Rerko RM, Rosser LE, Gately MK. Recombinant interleukin 12 cures mice infected with *Leishmania major.* J Exp Med 1993; 177:1505-1509.
56. Afonso LCC, Scharton TM, Vieira LQ, Wysocka M, Trinchieri G, Scott P. The adjuvant effect of interleukin-12 in a vaccine against *Leishmania major.* Science 1994; 263:235-237.
57. Macatonia SE, Hsieh C-S, Murphy KM, O'Garra A. Dendritic cells and macrophages are required for Th1 development of $CD4^+$ T cells from αβ TCR transgenic mice: IL-12 substitution for macrophages to stimulate IFN-γ production is IFN-γ-dependent. Internat Immunol 1993; 5:1119-1128.
58. D'Andrea AD, Aste-Amezaga M, Valainte NM, Ma X, Kubin M, Trinchieri G. Interleukin-10 inhibits lymphocyte IFN-γ production by suppressing natural killer cell stimulatory factor/interleukin-12 synthesis in accessory cells. J Exp Med 1993; 178:1041-1048.
59. Kubin M, Kamoun M, Trinchieri G. Interleukin-12 synergizes with B7/CD28 interaction in inducing efficient proliferation and cytokine production of human T cells. J Exp Med 1994; 180:211-222.
60. Murphy EE, Terres G, Macatonia SE, Hsieh C-S, Mattson J, Lanier L, Wysocka M, Trinchieri G, Murphy K, O'Garra A. B7 and interleukin-12 cooperate for proliferation and IFN-γ production by mouse Th1 clones that are unresponsive to B7 costimulation. J Exp Med 1994; 180:223-231.
61. Ding L, Linsley PS, Huang LY, Germain RN, Shevach EM. IL-12 inhibits macrophage costimulatory activity by selectively inhibiting the upregulation of B7 expression. J Immunol 1993; 151:1224-1234.
62. de Waal Malefyt R, Haanen J, Yssel H, Roncarolo M-G, te Velde A, Figdor C, Johnson K, Kastelein R, Spits H, de Vries JE. IL-10 and v-IL-10 strongly reduce antigen specific human T cell responses by diminishing the antigen presenting capacity of monocytes via down-regulation of class II MHC expression. J Exp Med 1991; 174:915-924.
63. Macatonia SE, Hosken NA, Litton M, Vieira P, Hsieh C-S, Culpepper J, Wysocka M, Trinchieri G, Murphy KM, O'Garra A. Dendritic cells produce IL-12 and direct the development of Th1 cells from naive CD4+ T cells. J Immunol 1995; 154:5071-5079.
64. Bradley LM, Duncan DD, Tonkonogy S, Swain SL. Characterization of antigen-specific CD4+ effector T cells in vivo: immunization results in a transient population of Mel-14^-, $CD45RB^-$ helper cells that secretes interleukin 2(IL-2), IL-3, IL-4 and interferon γ. J Exp Med 1991; 174:547-559.

65. Gollob KJ, Coffman RL. A minority subpopulation of CD4+ T cells directs the development of naive CD4+ T cells into IL-4 secreting cells. J Immunol 1994; 152:5180-5188.
66. Hosken NA, Shibuya K, Heath AW, Murphy KM, O'Garra A. The effect of antigen dose on CD4+ T cell phenotype development in an αβ-TCR-transgenic mouse model. J Exp Med 1995; 182:1579-1584.
67. Stern AS, Podlaski FJ, Hulmes JD, Pan Y-CE, Quinn PM, Wolitzky AG, Familletti PC, Stremlo DL, Truitt T, Chizzonite R, Gately MK. Purification to homogeneity and partial characterization of cytotoxic lymphocyte maturation factor from human B-lymphoblastoid cells. Proc Natl Acad Sci USA 1990; 87:6808-6812.
68. Hsieh C-S, Macatonia SE, O'Garra A, Murphy KM. Pathogen-induced Th1 phenotype development in CD4+ αβ-TCR transgenic T cells is macrophage dependent. Int Immunol 1993; 5:371-382.
69. Belosevic M, Finbloom DS, Meide PHVD, Slayter MV, Nacy CA. Administration of monoclonal anti-IFN-γ antibodies in vivo abrogates natural resistance of C3H/HeN mice to infection with *Leishmania major.* J Immunol 1989; 143:266-274.
70. Scott P. IFN-gamma modulates the early development of Th1 and Th2 responses in a murine model of Leishmaniasis. J Immunol 1991; 147:3149-3155.
71. Parronchi P, De Carli M, Manetti R, Simonelli C, Sampognaro S, Piccinni M-P, Macchia D, Maggi E, Del Prete G, Romagnani S. IL-4 and IFN (α and γ) exert opposite regulatory effects on the development of cytolytic potential by Th1 or Th2 human T cell clones. J Immunol 1992; 149:2977-2983.
72. Gollob KJ, Nagelkerken L, Coffman RL. Endogenous retroviral superantigen presentation by B cells induces the development of type 1 CD4+ T helper lymphocytes. Eur J Immunol 1993; 23:2565-2571.
73. Wang ZE, Reiner SL, Zheng S, Dalton DK, Locksley RM. CD4+ effector cells default to the Th2 pathway in interferon-γ-deficient mice infected with *Leishmania major.* J Exp Med 1994; 179:1367-1371.
74. Swihart K, Fruth U, Messmer N, Hug K, Behin R, Huang S, Del Guidice D, Aguet M, Louis JA. Mice from a genetically resistant background lacking the interferon γ receptor are susceptible to infection with Leishmania major but mount a polarized T helper cell 1-type CD4+ T cell response. J Exp Med 1995; 181:961-971.
75. Wenner CA, Macatonia SA, O'Garra A, Murphy KM. Common roles for IFNα and IFNγ in promoting Th1 differentiation induced by IL-12. J Exp Med 1995; in press:
76. Dighe AS, Campbell D, Hsieh C-S, Clarke S, Greaves DR, Gordon S, Murphy KM, Schreiber RD. Tissue-specific targeting of cytokine unresponsiveness in transgenic mice. Immunity 1995; 3:657-666.
77. Hsieh C-S, Macatonia SE, O'Garra A, Murphy KM. T cell genetic background determines default T helper phenotype development in vitro. J Exp Med 1995; 181:713-721.

78. Bendelac A, Schwartz RH. CD4+ and CD8+ T cells acquire specific lymphokine secretion potentials during thymic maturation. Nature 1991; 353:68-71.
79. Schmitt E, Hoehn P, Heuls C, Goedert S, Palm N, Rude E, Germann T. T helper type 1 development of naive CD4+ T cells requires the coordinate action of interleukin-12 and interferon-gamma and is inhibited by transforming growth factor-beta. Eur J Immunol 1994; 24:793-8.
80. Yanagida T, Kato T, Igarashi O, Inoue T, Nariuchi H. Second signal activity of IL-12 on the proliferation and IL-2R expression of T helper cell-1 clone. J Immunol 1994; 152:4919-4928.
81. Guler ML, Gorham JD, Hsieh C-S, Mackey AJ, Steen RG, Dietrich WF, Murphy KM. Genetic susceptibility to Leishmania: IL-12 responsiveness in T_H1 cell development. Science 1996; 271:984-987.
82. Chua AO, Chizzonite R, Desai BB, Truitt TP, Nunes P, Minetti LJ, Warrier RR, Presky DH, Levine JF, Gately MK, Gubler U. Expression cloning of a human IL-12 receptor component. A new member of the cytokine receptor superfamily with strong homology to gp130. J Immunol 1994; 153:128-136.
83. Chua AO, Wilkinson VL, Presky DH, Gubler U. Cloning and characterization of a mouse IL-12 receptor-b component. J Immunol 1995; 155:4286-4294.
84. Szabo SJ, Jacobson NG, Dighe AS, Gubler U, Murphy KM. Developmental commitment to the Th2 lineage by extinction of IL-12 signaling. Immunity 1995; 2:665-675.
85. Locksley RM, Scott P. Helper T-cell subsets in mouse Leishmaniasis: induction, expansion and effector function. Immunol Today 1991; 12.:58-61.
86. Howard JG, Hale C, Liew FY. Genetically determined susceptibility to leishmania tropica infection is expressed by hemopoietic donor cells in mouse radiation chimeras. Nature 1980; 288:161-162.
87. Swain SL, Huston G, Tonkonogy S, Weinberg AD. Transforming growth factor-beta and IL-4 cause helper T cell precursors to develop into distinct effector helper cells that differ in lymphokine secretion pattern and cell surface phenotype. J Immunol 1991; 147:2991.
88. Sad S, Mosmann TR. A single IL-2-secreting precursor CD4 T cell can develop into either a Th1 or Th2 secreting phenotype. J Immunol 1994; 53:3514-3522.
89. Hoehn P, Goedert S, Germann T, Koelsch S, Jin S, Palm N, Ruede E, Schmitt E. Opposing effects of TGF-beta 2 on the Th1 cell development of naive $CD4^+$ T cells isolated from different mouse strains. J Immunol 1995; 155:3788-93.
90. Nagelkerken L, Gollob K, Tielmans M, Coffman RL. Role of transforming growth factor β in the preferential induction of Th1 cells by staphyococcal enterotoxin. B Eur J Immunol 1993; 23:2306-10.
91. Barral-Netto M, Barral A, Brownell CE, Skeiky YAW, Ellingsworth LR, Twardzik DR, Reed SG. Transforming growth factor-b in Leishmanial

infection: a parasite escape mechanism. Science 1992; 257:545-548.
92. Miller A, Lider O, Roberts A, Sporn MB, Weiner HL. Suppressor T cells generated by oral tolerization to myelin basic protein suppress both in vitro and in vivo immune responses by the release of transforming growth factor β after antigen-specific triggering. Proc Natl Acad Sci USA 1992; 89:421-425.
93. Santambrogio L, Hochwald GM, Saxena B, Leu C-H, Martz JE, Carlino JA, Ruddle NH, Palladino MA, Gold LI, Thorbecke GJ. Studies on the mechanisms by which transforming growth factor-β (TGF-β) protects against allergic encephalomyelitis. J Immunol 1993; 151:1116-1127.
94. Oswald IP, Gazzinelli RT, Sher A, James SL. IL-10 synergizes with IL-4 and TGF-β to inhibit macrophage cytotoxic activity. J Immunol 1992; 148(11):3578-82.
95. Okamura H, Tsutsul H, Komatsu T, Yutsudo M, Hakura A, Tanimoto T, Torigoe K, Okura T, Mukuda Y, Hattori K, Akita K, Namba M, Tanabe K, Konishi K, Fukuda S, Kurimoto M. Cloning of a new cytokine that induces IFN-γ production by T cells. Nature 1995; 378:88-91.
96. Micallef M, Ohtsuki T, Kohno K, Tanabe F, Ushio S, Namba M, Tanimoto T, Torigoe K, Fujii M, Ikeda M, Fukuda S, Kurimoto M. Interferon-gamma inducing factor enhances T helper1 cytokine production by stimulated human T cells: synergism with IL-12 for interferon-gamma production. Eur J Immunol 1996; 26:1647-1651.
97. Bazan J, Timans J, Kastelein R. A newly defined interleukin-1? Nature 1996; 379:591.
98. Sadick MD, Heinzel FP, Holaday BJ, Pu RT, Dawkins RS, Locksley RM. Cure of leishmaniasis with anti-interleukin 4 monoclonal antibody. Evidence for a T-cell-dependent, interferon gamma-independent mechanism. J Exp Med 1990; 171:115-127.
99. Chatelain R, Varkila K, Coffman RL. IL-4 induces a Th2 response in *Leishmania major*-infected mice. J Immunol 1992; 148:1182-1187.
100. Romani L, Mencacci A, Grohmann U, Mocci S, Mosci P, Puccetti P, Bistoni F. Neutralizing antibody to interleukin 4 induces systemic protection and T helper type 1-associated immunity in murine candidiasis. J Exp Med 1992; 176:19.
101. Kopf M, Le Gros G, Bachmann M, Lamers MC, Bluethmann H, Kohler G. Disruption of the murine IL-4 gene blocks Th2 cytokine responses. Nature 1993; 362:245-248.
102. Noben-Trauth N, Kropf P, Muller I. Susceptibility to Leishmania major infection in interleukin-4-deficient mice. Science 1996; 271:987-990.
103. Constant S, Pfeiffer C, Woodard A, Pasqualini T, Bottomly K. Extent of T cell receptor ligation can determine the functional differentiation of naive $CD4^+$ T cells. J Exp Med 1995; 182:1591-1596.
104. Fowell D, McKnight AJ, Powrie F, Dyke R, Mason D. Subsets of CD4+ T cells and their roles in the induction and prevention of autoimmunity. Immunol Rev 1991; 123:37-64.

105. Steinman L. Escape from "horror autoxicus": pathogenesis and treatment of autoimmune disease. 1995; Cell 80:7-10.
106. Liblau R, Singer S, McDevitt H. Th1 and Th2 CD4+ T cells in the pathogenesis of organ-specific autoimmune diseases. Immunol Today 1995; 16:34-38.
107. Mocci S, Coffman RL. The induction of a Th2 population from a polarized Leishmania-specific Th1 population by in vitro culture. J Immunol 1995; 154:3779-3787.
108. Perez VL, Lederer JA, Lichtman AH, Abbas AK. Stability of Th1 and Th2 populations. Inter Immunol 1995; 7:869-875.
109. Murphy E, Shibuya KK, Hosken N, Openshaw P, Maino V, Davis K, Murphy K, O'Garra A. Conversion of T Helper 1 and T Helper 2 Cytokine Profiles in antigen-specific CD4+ T cell populations. J Exp Med 1996; 183:901-913.
110. Romani LA, Mencacci U, Grohmann SM, Mosci P, Puccetti P, Bistoni F. Neutralizing antibody to interleukin 4 induces systemic protection and T helper type 1-associated immunity in murine candidiasis. J Exp Med 1992; 176:19.
111. Nabors GS, Afonso LCC, Farrell JP, Scott P. Switch from a type 2 to a type 1 T helper cell response and cure of established *Leishmania major* infection in mice is induced by combined therapy with interleukin 12 and pentostam. Proc Natl Acad Sci USA 1995; 92:3142-3146.
112. Correa-Oliveira R, Varkila K, Coffman RL. The adoptive transfer of Leishmania-specific Th1 and Th2 responses can be regulated by IL-4 and IFN-γ. Eur J Immunol 1995; in press:
113. Holaday BJ, Sadick MD, Wang ZE, Reiner SL, Heinzel FP, Parslow TG, Locksley RM. Reconstitution of *Leishmania* immunity in severe combined immunodeficient mice using Th1- and Th2-like cell lines. J Immunol 1991; 147:1653-1658.
114. Manetti R, Gerosa F, Giudici MG, Biagotti R, Parronchi P, Piccinni M-P, Sampognaro S, Maggi E, Romagnani S, Trinchieri G. Interleukin-12 induces stable priming for interferon-γ (IFN-γ) production during differentiation of human T helper (Th) cells and transient IFN-γ production in established Th2 cell clones. J Exp Med 1994; 179:1273-1283.

CHAPTER 2

Heterogeneity of Th1 Type Responses to Type II Collagen—Implications for the Development of Arthritis

Tieno Germann

INTRODUCTION

Type II collagen (CII)-induced arthritis is an autoimmune joint disease of rodents resembling human rheumatoid arthritis. Induction of the disease in susceptible DBA/1 mice requires immunization with CII emulsified with complete Freund's adjuvant (CFA). Immunization with CII in incomplete Freund's adjuvant (IFA) results in an intermediate CII-specific immune response and a low incidence of arthritis (0-20%). We found that IL-12 can profoundly enhance the cellular and humoral immune reaction to CII and trigger severe destructive arthritis with an incidence of 80-100% in CII in IFA-immunized DBA/1 mice. In contrast, IL-12 failed to upregulate the CII-specific humoral immune reaction in allogenic or congenic mice with C56BL background immunized with CII in IFA. Despite promoting a strong CII-specific Th1 cell response at the T cell level in mice with C56BL background, injections of IL-12 did not induce joint disease in these mice. Likewise, neutralization of IFN-γ in IL-12-treated DBA/1 mice immunized with CII in IFA completely prevented arthritis. This treatment resulted in reduced serum levels of CII-specific (IgG2a) antibodies but rather enhanced development of CII-specific Th1 cells. These results point to the important pathogenic role of IgG2a autoantibodies in CII-induced arthritis. By administration of IL-12 antagonists to DBA/1 mice immunized with CII in CFA, we provide evidence that the adjuvant effect of the mycobacterial components contained in CFA might at least partially be due to

Immunointervention in Autoimmunity by Th1/Th2 Regulation, edited by Luciano Adorini.

the induction of endogenous IL-12 synthesis. Finally, we observed that high doses of IL-12 applied to DBA/1 mice immunized with CII in CFA can mitigate the course of joint disease. Under these conditions, IL-12 suppressed the humoral immune reaction to CII, with most pronounced effects on IgG1 synthesis. Our data will be discussed in the context of the regulation of humoral immune responses by IL-12 and IFN-γ and the consequences of CII-specific antibody synthesis of different Ig isotypes for the development of arthritis.

LYMPHOCYTE SUBSETS AND IFN-γ IN AUTOIMMUNE DISEASE

T Cells and B Cells/Antibodies in Autoimmune Diseases

Most autoimmune diseases have a complex pathogenesis determined by MHC genes, largely unidentified non-MHC background genes, and environmental influences.[1-3] The interplay between all these factors results in the development of autoimmunity. During the course of many autoimmune diseases, reactivity against self (proteins) can be detected at the level of T and B cells. In some cases, autoantibodies directly trigger the disease. A well-known example is acquired hemolytic anemia in mice[4] and humans[5] where autoantibodies to erythrocytes cause the destruction of these cells resulting in anemia. Likewise, autoantibodies to the thyroid-stimulating hormone receptor, to the surface of epidermal skin cells (to the desmoglein I glycoprotein), and to the acethylcholine receptor are responsible for the development of Graves' disease, Pemphigus vulgaris, and myasthenia gravis, respectively[5]. Efficient (auto)antibody production requires T cell help. Thus, self-reactive T cells are of crucial importance in the pathogenesis of autoimmunity. In many cases, disease-promoting T cells belong to the CD4-positive subset. Their participation is also reflected by the association of several murine and human autoimmune disorders with certain MHC class II alleles.[1] Recently, Liblau et al[6] suggested that the differentiation of autoreactive CD4-positive T cells into Th1 type cells might be a prerequisite for the development of organ-specific autoimmune diseases such as experimental autoimmune encephalomyelitis (EAE)/multiple sclerosis (MS) and insulin-dependent diabetes mellitus (IDDM) in humans and mice.

Differentiation of Naive CD4-Positive Th Cells into Th1 and Th2 Cells and Their Role in the Regulation of Humoral Immunity and in the Pathogenesis of Autoimmunity

Naive CD4-positive T cells can develop into different effector cells. This process is influenced by several parameters. Among those, the avidity of the interaction between the T cell receptor and its peptide/MHC ligand on the surface of the APC,[7-9] as well as cytokines,[10,11] appear to be of crucial importance. So-called Th1 and Th2 cells probably represent the extremes of a broad spectrum of functionally different Th cells. They were first described in the mouse by the groups of T. Mosmann[12] and K. Bottomly.[13]

Both Th cell subsets are defined on the basis of the cytokines secreted upon stimulation and the different effector functions exerted in vivo. Th1 cells selectively secrete interferon–γ (IFN-γ), interleukin-2 (IL-2), and tumor necrosis factor-β (TNF-β). They mediate delayed type hypersensitivity responses[14,15] and are potent activators of macrophages.[16] Both functions are associated with classical "cellular immunity", which is responsible for the protection against intracellular pathogens.[17] Conversely, Th2 cells, which exclusively secrete the cytokines IL-4 and IL-5, are considered to mediate "humoral immunity" due to their particularly good helper function for B cells. They primarily promote antibody synthesis of the IgE and IgG1 isotypes and support the proliferation of mast cells and eosinophil granulocytes.[12,18,19] Cellular and humoral immunity are regulated in a separate and often mutually exclusive way.[20,21] In mice mounting a good proliferative response and IFN-γ synthesis but little antibody production in response to collagen type IV, neutralization of IFN-γ in vivo resulted in the development of Th2 cells, enhanced synthesis of IL-5, and formation of large amounts of specific IgG1 antibodies.[21] This shows that IFN-γ under certain conditions can inhibit humoral immunity. However, some Th1 clones supported antibody synthesis, especially of the IgG2a subclass, in vitro[18,19] and in vivo.[22,23] Likewise, administration of IFN-γ can upregulate IgG2a synthesis in vivo.[24] In agreement with a role of IFN-γ in the upregulation of IgG2a production, the serum levels of hapten-specific IgG2a antibodies were about 10-fold reduced in IFN-γ receptor-deficient mice immunized with a hapten-protein conjugate, whereas the serum levels of total Ig, IgM, IgG1and IgG2b were equivalent in mutant and wild-type mice.[25] The effects of IFN-γ on antibody synthesis can be dose-dependent. Whereas low doses of IFN-γ enhanced the synthesis of IgG2a[24,26] and in combination with IL-5 that of IgG3 antibodies,[26] IFN-γ, at higher doses, inhibited antibody production, in particular that of IgG1 and IgE.[24,26] Taken together, these results clearly indicate that the classification of Th1 cells as mediators of "cellular" immunity represents an oversimplification. At least some Th1 cells can promote both "cellular" and also "humoral" immunity.

The most important cytokines regulating Th cell development are IL-4 and IL-12. IL-4 drives the differentiation of naive CD4-positive cells in a Th2 direction,[27,28] whereas IL-12 promotes Th1 development.[10,11] IL-12 is a recently cloned heterodimeric cytokine, composed of covalently linked p35 and p40 chains. It affects proliferation, cytokine (IFN-γ) production and cytotoxicity of CD4 and CD8-positive T cells as well as NK cells. IL-12 is primarily produced by monocytes/ macrophages and dendritic cells after stimulation with microbial products or in the interaction with activated Th cells (reviewed in refs. 29, 30).

There is increasing evidence that the functional dichotomy of Th cells is of relevance for understanding the pathogenesis of organ/tissue-specific autoimmune diseases such as IDDM or EAE as reviewed in ref. 6. In agreement with the disease-promoting function of Th1 type cells and the role of IL-12 as the major cytokine inducing Th1 cell development, administration of IL-12 or inhibitors of IL-12 enhanced or mitigated the course of several autoimmune diseases such as diabetes (IDDM), EAE, arthritis, and granulomatous colitis, respectively (reviewed in refs. 30, 31). In contrast, administration of IL-4 suppressed autoimmunity in several models (reviewed in refs. 6, 31). Treatment with IL-4 also facilitated remissions in the CII-induced arthritis model.[32] Furthermore, the incidence and severity of this disease were reduced in animals inoculated with IL-4 or IL-13-transfected fibroblasts[33] and daily injections of IL-4 ameliorated the chronic, destructive phase of streptococcal cell wall-induced arthritis.[34] Besides the ability of IL-4 to modulate the development of (auto)antigen-specific T cells by inducing Th2 cells, Th2 type cytokines such as IL-4 and IL-10 may suppress autoimmunity by downregulating the production of proinflammatory cytokines and the induction of anti-inflammatory molecules such as the IL-1 receptor antagonist by monocytes/macrophages.[35,36] This type of immunoregulation by IL-4 and IL-10 has also been demonstrated using synovial cell cultures from patients with rheumatoid arthritis.[37,38]

The Role of IFN-γ in Immune Responses and Autoimmune Diseases

IFN-γ represents the key cytokine produced by Th1 cells. It is a potent immunoregulatory factor which plays an important role in the activation of macrophages, in regulating antigen presentation, and in modulating antibody responses.[39]

IFN-γ is obligatory in the defense against intracellular microorganisms. Mice with a targeted disruption of the IFN-γ gene were unable to control the growth of a normally sublethal dose of *Mycobacterium tuberculosis*.[40] IFN-γ is also critically involved in the control of many viral infections. Most viral infections elicit strong humoral immune responses. The antiviral antibodies found during infections with lactate dehydrogenase-elevating virus as well as adenovirus were predominantly of the IgG2a subclass[41] and their production depended on the presence of IFN-γ-producing CD4-positive T cells.[42] Influenza virus-specific CD4-positive T cells mediated recovery from pulmonary infection in the presence of B cells.[43] A second study demonstrated that CD4-positive Th1 but not Th2 cells are protective in influenza infection.[44] Likewise, poliovirus-specific Th1 cells in combination with primed B cells protected mice against lethal poliovirus infection in transgenic animals expressing the human poliovirus receptor.[23] These MHC-class II-restricted CD4-positive Th1 cells/clones were cytotoxic against poliovirus peptide-pulsed target cells in vitro and promoted the synthesis of neutralizing poliovirus-specific antibodies in vivo. The dominant antibody subclasses produced were IgG2a and to a lesser extent IgG2b. In agreement with the assumption that IFN-γ produced by Th1 cells regulates the production of neutralizing virus-specific IgG2a antibodies, IFN-γ receptor-deficient 129/Sv/Ev mice, despite mounting a Th1 type response at the T cell level after inoculation with live attenuated pseudorabies virus, did not resist a challenge infection with pseudorabies virus, probably because of the about 10-fold reduction in virus-specific IgG2a production compared to wild-type littermates.[45] Thus, the effects of IFN-γ on B cells/antibody synthesis appear to mediate protective immunity in this viral disease model. Furthermore, mice with a targeted disruption of the IFN-γ receptor gene succumbed to infection with vaccinia virus. The animals were unable to restrict the growth of vaccinia virus within the first few days of infection and died despite a normal response of cytotoxic T lymphocytes.[25] In contrast, the course of infection of such mice with vesicular stomatitis virus was unaltered and the titers of neutralizing antibodies were comparable to those found in control animals.[25] Likewise, mice with a targeted disruption of the IFN-γ gene developed an effective humoral and cytotoxic T cell response to infection with influenza virus. No difference in the time to death and minimal lethal dose was found between infected wild-type and IFN-γ knockout animals. Analysis of the subclass pattern of influenza virus-specific antibodies revealed only slightly decreased IgG2a titers and increased IgG1 titers in IFN-γ knockout mice compared to wild-type mice.[46] Thus, the production of virus-specific IgG2a antibodies is IFN-γ-dependent in some but not all models of viral infections. This indicates that IFN–γ can enhance but is not essential for IgG2a synthesis, a conclusion that can also be drawn from studies with IFN-γ receptor-knockout mice.[25] In those viral models were IFN-γ receptor knockout mice mounted a protective immune response, other cytokines such as IFN-α[47] may substitute for IFN-γ in promoting the synthesis of neutralizing IgG(2a) antibodies. Alternatively, protection might be mediated solely by CD8 and/or CD4-positive cytotoxic T cells destroying virally-infected cells.

Besides the induction of MHC class II expression on professional and non-professional APC, the enhancing effect of IFN-γ on the synthesis of IgG2a antibodies may also be of relevance for the disease-promoting effect of this cytokine in some autoimmunity models. Transgenic expression of IFN-γ in the neuromuscular junctions of mice resulted in a myasthenia gravis-like disease.[48] A strong humoral autoimmune response towards components of the motor end plates with focal deposition of IgG in synaptic clefts was observed. Muscles were infiltrated mainly by macrophages and some CD4-positive T cells. The adjacent lymph nodes contained plenty of plasma cells. This suggests that autoantibody production towards the neuromuscular junctions with activation of macrophages caused this experimental disease. IFN-γ seems to be involved in the pathogenesis of diabetes in NOD mice since early treatment of NOD mice with a neutralizing antibody against IFN-γ reduced the incidence of diabetes and decreased the severity of insulitis.[49,50] Furthermore, transgenic expression of IFN-γ in pancreatic B cells induced an autoimmune response towards pancreatic tissue. However, in this model no evidence for a humoral autoimmune response towards pancreatic tissue was observed. The destruction of pancreatic tissue was mediated by CD8-positive cytotoxic T lymphocytes.[51]

In contrast to NOD mice, neutralization of IFN-γ exacerbated the course of disease in other models of autoimmunity. In C57BL/6 mice, which are normally relatively resistant to the induction of EAE, injection of an anti-IFN-γ mAb led to an increase in disease morbidity and resulted in a mortality of up to 80%. Conversely, administration of exogenous IFN-γ to highly susceptible SJL/J mice reduced morbidity and mortality.[52] That IFN-γ is not required for the development of EAE was demonstrated in IFN-γ knockout mice. Regarding the induction and the clinical course of EAE as well as the inflammatory infiltrate within the central nervous system, no difference was observed between IFN-γ knockout mice, heterozygotic mice, and wild-type littermates.[53] Furthermore, in the model of experimental autoimmune uveoretinitis, neutralization of endogenous IFN-γ resulted in a much more severe disease in susceptible mice and allowed the expression of this disease in several, otherwise resistant, strains of mice.[54] Anti-IFN-γ-treated mice displayed augmented delayed hypersensitivity responses to the autoantigen used for immunization and an increase in monocyte/macrophage infiltration in the ocular lesions. Taken together, these results show that IFN-γ can mitigate EAE and experimental autoimmune uveoretinitis and that IFN-γ is not obligatory in the pathogenesis of at least one organ specific autoimmune disease, namely EAE.

THE TYPE II COLLAGEN (CII)-INDUCED ARTHRITIS MODEL

Type II collagen (CII)-induced arthritis is a tissue-specific autoimmune disease developing in susceptible strains of rat[55] and mouse.[56] The autoantigen CII is a major component of joint cartilage. The susceptibility to CII-induced arthritis is associated with certain MHC class II alleles. Mice bearing the H-2^q haplotype (DBA/1, NFR/N) are highly susceptible to disease, whereas animals bearing the H-2^b haplotype (C57BL/6, C57BL/10) are not.[57,58] T and B lymphocytes are activated during the immune reaction to CII and the development of full-blown joint disease probably relies on cellular as well as humoral immune reactions towards CII. CD4-positive T cells play a crucial role in the pathogenesis of the disease. This is indicated by the prevention of arthritis in CII in CFA-immunized DBA/1 mice injected with an anti-CD4 mAb.[59] The CII-specific T cells in DBA/1 mice appear to belong to the Th1 subset of CD4-positive T lymphocytes. Activation of spleen cells from animals immunized with CII in

CFA with denatured CII in vitro resulted in the synthesis of IFN-γ.[60] However, arthritis was not observed if the humoral anti-CII response was blocked,[61,62] indicating that CII-specific antibodies also contribute to the development of joint disease. A high antibody response towards CII was observed in susceptible strains of mice. However, it was not the overall CII-specific antibody production but rather the CII-specific IgG2a (and to some degree IgG2b) response that appeared to contribute to the pathogenesis of CII-induced arthritis.[57,63]

CII-induced arthritis can be described clinically and histologically. Clinically, the disease is characterized by erythema and swelling of numerous peripheral joints. Isolated joints or whole feet (from the ankle to the distal interphalangeal joints) can be affected. This is reflected in the clinical scoring system utilized by our group where 0 indicates: no disease (normal appearance), 1: swelling of one or two interphalangeal joints, 2: swelling of more than two interphalangeal joints or slight swelling of the ankle region with erythema of the foot, 3: massive swelling of the ankle and metatarsus (-carpus), and 4: massive swelling of the entire paw (including the ankle and the distal interphalangeal joints), ankylosis, limited mobility. At later stages of CII-induced arthritis, affected mice suffer from severely destroyed joints and ankylosis (score 3 or 4) in about 50% of their footpaws. Histologically, edema and cellular infiltration of the synovium, the joint cavity, and the surrounding tissue can be observed. Initially, the inflammatory infiltrate consists mainly of polymorphonuclear and subsequently of mononuclear cells. Synovial hyperplasia (pannus formation) develops, resulting in erosion of cartilage and bone as well as new bone formation. This frequently causes destruction of joint structures or ankylosis of joints.[55-57,63]

In many clinical and histological aspects CII-induced arthritis in rats or mice resembles human rheumatoid arthritis[64] and serves as a relevant experimental animal model of this human disease. CII-specific immune reactivity at the cellular and humoral level[64,65] can be demonstrated in patients and CII may be one of the autoantigens involved in the pathogenesis of rheumatoid arthritis. In light of the evidence that arthritis might be at least partially a Th1-mediated disease, we decided to study the effects of IL-12, the so far most potent promoter of Th1 type immune responses, in the CII-induced arthritis model. The following questions were addressed specifically:

- can exogenous IL-12 replace bacterial products as adjuvant and promote joint disease in DBA/1 mice immunized with CII in IFA? How does IL-12 affect the cellular and humoral immune response to CII?
- is endogenous IL-12 involved in the adjuvant effect of bacterial products?
- what is the role of IL-12-induced IFN-γ regarding the adjuvant effect of IL-12?

-is IL-12 treatment able to break the resistance of C57BL/6 mice to CII-induced arthritis? What is the influence of IL-12 on the CII-specific immune response in C57BL/6 mice?
-is there any effect of IL-12 on the CII-specific immune response and the course of arthritis when administered in combination with bacterial adjuvant to CII-immunized DBA/1 mice?

RESULTS OF IL-12 TREATMENT IN MICE

Injections of IL-12 Promote the Development of a Severe Destructive Arthritis with High Incidence in DBA/1 Mice Immunized with CII in IFA

Arthritis occurs with a low incidence of 10-20% in DBA/1 mice immunized with CII in IFA intradermally and boosted with CII in PBS intraperitoneally 3 weeks later. The disease of these few arthritic mice is usually restricted to single joints/digits and not associated with a strong inflammatory response/swelling of the whole paw. When DBA/1 mice injected with CII in IFA were additionally treated with IL-12 for 5 days (day -1 to +3) simultaneously with the priming and booster immunizations, a strongly increased incidence of arthritis was observed. In more than 10 experiments with at least 10 mice per group 80-100% of the animals treated in this way became arthritic.[66] Besides increasing the incidence of arthritis, administration of IL-12 also enhanced the severity of the disease, as characterized by measurements of footpad and ankle thickness as well as histological studies. About 50% of the limbs of CII in IFA-immunized plus IL-12-treated DBA/1 mice showed erythema and massive swelling of their paws. The swelling typically included the ankle region, the metatarsus/carpus and the digits. Histological examination of longitudinal sections of paws prepared 40-60 days after priming revealed an extensive inflammatory reaction comprising several joints and the surrounding tissue. Fibrin exudation into the joint cavity was observed. The synovial lining cells exhibited hypertrophic and hyperplastic changes (pannus formation) and destruction of cartilage surfaces and erosion of bones occurred. These pathological changes frequently resulted in the loss of the joint structure.

IL-12 Treatment of DBA/1 Mice Immunized with CII in IFA Promotes a CII-Specific Humoral and Cellular Th1 Type Autoimmune Response

In order to characterize the effects of IL-12 treatment on the immune reaction in DBA/1 mice immunized with CII in IFA plus/minus IL-12, all mice were bled at day 28 after priming and spleen cells as well as serum were prepared at the end of the respective experiments (day 80 -120 after priming). The humoral immune response was characterized by determination of the titers of CII-specific total immunoglobulin as well as IgG subclasses in the sera of DBA/1 mice. The cellular immune response to CII in DBA/1 mice was investigated by stimulating spleen cells ex vivo with denatured CII and measuring the synthesis of IFN-γ after 24, 48 and 72 hours.

IFN-γ synthesis in the absence of CII in vitro was below 10 U/ml in all cultures. Upon CII-specific activation in vitro, spleen cells from mice immunized with CII in IFA produced rather low amounts (5-20 U/ml) of IFN-γ. Administration of IL-12 in vivo during the immunizations with CII in IFA promoted the development of splenic T cells producing IFN-γ in DBA/1 mice of both sexes. The production of IFN-γ by spleen cells activated with denatured CII ex vivo was elevated about 3- to 10-fold. IFN-γ was almost exclusively produced by $CD4^+$ T cells because depletion of $CD4^+$ but not $CD8^+$ T cells prior to antigen specific activation of spleen cells eliminated the synthesis of IFN-γ.[66] Thus, IL-12 induced $CD4^+$ T cells of the Th1 type in DBA/1 mice immunized with CII in IFA.

Administration of IL-12 also resulted in a pronounced upregulation of the CII-specific humoral immune reaction in DBA/1 mice. Immunization of DBA/1 mice with CII in IFA generated an intermediate antibody response which was dominated by antibodies of the IgG1 subclass as well as some IgG2b and IgG2a antibodies. Treatment of mice with IL-12 in vivo enhanced the synthesis of CII-specific total Ig 3 to 10-fold. Whereas antibodies of the IgG1 subclass were upregulated only slightly (about 2-fold), CII-specific antibodies of the IgG2b and IgG2a subclasses were upregulated 3 to 10-fold and 10 to 50-fold, respectively.[66-68] This demonstrates that IL-12 treatment strongly augments the production of those IgG subclasses (IgG2a and IgG2b) that are critical in the pathogenesis of CII-induced arthritis. Taken together, these results clearly document the potent adjuvant effect of IL-12 on the cellular and humoral CII-specific immune reaction in DBA/1 mice.

The IL-12-Induced Increase of CII-Specific Antibody Production and the Development of Joint Disease Are IFN-γ-Dependent

One of the major effects of IL-12 is the upregulation of IFN-γ production by T and NK cells.[29,30] This led us to investigate whether or not the effects of IL-12 in the CII-induced arthritis model are IFN-γ-dependent. In two independent experiments DBA/1 mice were immunized with CII in IFA and simultaneously injected with medium (as a control), IL-12, or IL-12 in combination with a neutralizing mAb to IFN-γ. Only a few animals of the control group developed signs of joint disease (Exp.1: 2/10, Exp.2: 1/4). Again, IL-12 treatment resulted in a strong increase in the incidence of arthritis (Exp.1: 10/10, Exp.2: 7/7). However, when CII in IFA-immunized animals received IL-12 in combination with an anti-IFN-γ mAb none of the animals showed signs of arthritis (Exp.1: 0/10, Exp.2: 0/5). Thus, the promotion of autoimmune arthritis by IL-12 depends on the endogenous synthesis of (IL-12-induced) IFN-γ. Neutralization of IFN-γ in vivo also affected the immune reaction to CII. Determination of the CII-specific IgG2a response revealed an enormous increase in the sera of IL-12-treated mice (Exp.1: 30-fold, Exp.2: 70-fold) compared to CII in IFA-immunized mice. This increase was almost completely (> 90%) prevented by neutralization of IFN-γ. Surprisingly, the development of CII-specific Th1 type cells was not impaired in IL-12 plus anti-IFN-γ-treated mice. The production of IFN-γ by spleen cells of these mice activated in vitro with denatured CII was rather enhanced (about 3-fold) as compared to spleen cells of CII in IFA-immunized and IL-12-treated mice. This indicates that the mere development of a marked Th1 type reaction at the level of T cells alone (in combination with a low to intermediate CII-specific antibody response) is not sufficient to cause arthritis. Alternatively, the Th1 cells developing in CII in IFA-immunized DBA/1 mice injected with IL-12 might be functionally different from those developing in IL-12 plus anti-IFN-γ-treated animals.

IL-12 Upregulates CII-Specific Antibody Synthesis and Promotes Joint Disease Only in DBA/1 but Not in C57BL/6 or BALB/c Mice Immunized with CII in IFA

CII-specific immune responses can be elicited in several strains of mice such as DBA/1, C57BL/6 or BALB/c.[57] When mice of these strains were immunized with CII in IFA plus/minus IL-12 treatment, we noted[67] that IL-12 promoted clinical signs of arthritis (erythema, swelling) only in DBA/1 mice (9/10) but not in C57BL/6 (0/10) or BALB/c (0/10). Furthermore, histological examination of 10 nonarthritic hindpaws of IL-12-treated C57BL/6 mice did not show signs of disease (no infiltration by inflammatory cells, no thickening/ proliferation of the synovial lining cells, no erosion of cartilage or bone).

Characterization of the humoral immune responses revealed that the titers of total CII-specific antibodies reached comparable (intermediate) levels in DBA/1 and C57BL/6 mice immunized with CII in IFA, whereas those found in BALB/c mice were about 10-fold lower. Interestingly, administration of IL-12 upregulated (10-fold) antibody synthesis to high serum levels only in DBA/1 but not in C57BL/6 or BALB/c mice.[67,68] This again indicates that the induction of CII-induced arthritis with IL-12 as adjuvant correlates with a pronounced upregulation of the CII-specific humoral immune response. When characterized in detail, the most striking effect of IL-12 treatment was observed on the synthesis of the IgG2a and IgG2b subclasses. Whereas a tremendous increase in the serum levels of CII-specific IgG2a and IgG2b was noted in IL-12-treated DBA/1 mice compared to DBA/1 mice immunized with CII in IFA alone, injections of IL-12 failed to upregulate (but instead slightly reduced in most experiments) the CII-specific IgG2a and IgG2b serum levels in C57BL/6 mice.

In contrast to the humoral CII-specific immune reaction, IL-12 promoted the generation of IFN-γ-producing T cells equally well in DBA/1 and C57BL/6 mice.[68] Single cell suspensions of spleens prepared from DBA/1 and C57BL/6 mice immunized with CII in IFA plus/minus IL-12 treatment were cultured in vitro in the absence or presence of CII for 48 hours. Thereafter, activation supernatants were collected and the synthesis of IFN-γ determined by ELISA. In the absence of CII only low amounts of IFN-γ were produced. Addition of CII induced some IFN-γsynthesis in the cultures of spleen cells from either DBA/1 or C57BL/6 mice immunized with CII in IFA alone. About 10-fold higher levels of IFN–γ were present in the supernatants of spleen cell cultures of both DBA/1 and C57BL/6 mice immunized with CII in IFA and treated with IL-12 in vivo. Thus, the effect of IL-12 on the development of CII-specific Th1 type cells is comparable in DBA/1 and C57BL/6 mice.

In (DBA/1xC57Bl/6) F1 Mice Immunized with CII in IFA, Administration of IL-12 Promotes Arthritis and Upregulates CII-Specific IgG2a And IgG2b Production

We next investigated whether the relative resistance of C57BL/6 mice or the susceptibility of DBA/1 mice to the induction of CII-induced arthritis with IL-12 as adjuvant is dominant. Therefore, we generated (DBA/1xC57BL/6) F1 mice which were immunized with CII in IFA plus/minus IL-12 treatment.[68] Injections of IL-12 at several doses accelerated the onset and increased the incidence of joint disease. This was accompanied by a strong increase in CII-specific IgG2a and IgG2b serum levels as well as in enhanced IFN–γ synthesis by spleen cells stimulated ex vivo with CII as antigen. Thus, the susceptibility of DBA/1 mice regarding the effects of IL-12 on the synthesis of CII-specific antibodies and the development of joint disease are dominant in (DBA/1xC57BL/6) F1 mice.

Non-MHC Background Genes Are Probably Responsible for the Failure of C57BL Mice to Upregulate CII-Specific Antibody Synthesis and to Develop Arthritis when Injected with IL-12

C57BL/6 mice differ from DBA/1 mice regarding the MHC haplotype (H-2^b versus H-2q) as well as to their non-MHC genetic background. This raised the question of whether MHC or non-MHC genes were responsible for the converse effect of IL-12 on the CII-specific antibody response and the appearance of joint disease.

We therefore immunized C57BL/10.Q mice, which express the same MHC haplotype as DBA/1 mice, with CII in IFA and injected the animals with medium or various doses of IL-12 (1000, 200, 40 ng/day). Similarly to C57BL/6 mice, administration of IL-12 to CII in IFA-immunized C57BL/10.Q mice was unable to induce clinical signs of joint disease. Injections of IL-12 were associated with slightly downregulated CII-specific IgG2a and IgG2b antibody levels as compared to the intermediate serum levels detected in the sera of C57BL/10.Q mice immunized with CII in IFA. The differential effect of IL-12 treatment in DBA/1 and C57BL/6 and C57BL/10.Q mice is highly reproducible. The failure of IL-12 to promote arthritis and to enhance CII-specific antibody synthesis in C57BL/6 or C57BL/10.Q mice was observed in each of more than 15 independent experiments. At the cellular level, IL-12 promoted the generation of IFN-γ-producing Th1 type cells in C57BL/10.Q mice as well, although less pronounced than in C57BL/6 or DBA/1 mice.[68] These results suggest that the non-MHC genetic background of C57BL/10.Q and C57BL/6 mice is probably responsible for the failure of IL-12 to upregulate the CII-specific antibody response and to promote arthritis in these strains of mice.

Anti-IL-12 Treatment Weakly Suppressed IgG2a Synthesis and Inconsistently Ameliorated Joint Disease in DBA/1 Mice Immunized with CII in CFA

Having shown that IL-12 is a potent adjuvant for triggering autoimmune arthritis in DBA/1 mice and bearing in mind our assumption that the adjuvant effect of mycobacteria might be due to the induction of endogenous IL-12, we next investigated the effects of anti-IL-12 reagents on the course of arthritis and the CII-specific immune reaction in DBA/1 mice immunized with CII in CFA.[67] CFA contains killed mycobacteria as adjuvant and immunization of DBA/1 mice with CII in CFA elicits a strong autoimmune response with high serum titers of CII-specific antibodies and severe destructive arthritis. The anti-IL-12 reagents employed were the IL-12p40 homodimer (continuous daily treatment for 5 weeks) either alone or in combination with an IL-12 heterodimer-specific mAb. The anti-IL-12 treatment mitigated the course of joint disease and reduced the serum levels of CII-specific IgG2a antibodies slightly in two of five experiments.[67] However, in other experiments, anti-IL-12 treatment had either no effect at all, or, the injection of control reagents also suppressed the disease. Thus, although exogenous IL-12 can replace mycobacteria as adjuvant and can induce joint disease in DBA/1 mice immunized with CII in IFA, the role of endogenous IL-12 in the development of arthritis in DBA/1 mice immunized with CII in CFA needs further clarification.

Treatment of DBA/1 Mice Immunized with CII in CFA with High Doses of IL-12 Reduces the Incidence and Severity of Joint Disease

The common way to induce arthritis in DBA/1 mice depends on the immunization with CII emulsified with complete Freund's adjuvant (CFA), where killed mycobycteria serve as adjuvant. When DBA/1 mice immunized with CII in CFA were simultaneously treated with high doses of IL-12, we made the unexpected observation of a mitigation of the disease. First, high dose IL-12 treatment reduced the incidence of arthritis from 62% to 21%, 76% to 33%, 93% to 56% in three independent experiments.[69] Second, the disease was less severe in the mice treated with IL-12 that were arthritic. In most cases restriction to digits and only occasionally swelling of entire feet were noted. Thus, IL-12 can reduce the incidence and severity of CII-induced arthritis when administered together with CFA (killed mycobacteria). Injections of high doses of IL-12 (1 μg/day, 5 days/week) for at least two weeks (starting one day prior to priming) are necessary to successfully mitigate arthritis. Ten-fold lower doses of IL-12 did not exert a comparable effect. When IL-12 treatment was delayed and started three weeks after priming at the time of the intraperitoneal booster immunization with CII in PBS (shortly before arthritis symptoms show up), the appearance and progression of the disease were not influenced.

Characterization of the CII-specific immune reaction in DBA/1 mice immunized with CII in CFA plus/minus treatment with high doses of IL-12 was performed at the cellular and humoral level. The production of IFN-γ by spleen cells prepared from animals of both groups of mice and activated in vitro with denatured CII was equivalent. However, the humoral CII-specific immune response was modified in a characteristic way. DBA/1 mice immunized with CII in CFA alone showed a marked CII-specific humoral immune response (which is equivalent in magnitude and IgG subclass distribution to the humoral response observed in DBA/1 mice immunized with CII in IFA plus IL-12). Large amounts of CII-specific IgG1, IgG2a and IgG2b antibodies can be detected in the sera of such animals. The additional treatment of DBA/1 mice immunized with CII in CFA with high doses of IL-12 downregulated the CII-specific antibody response.[69] The most pronounced effect was observed for antibodies of the IgG1 subclass. Their serum titers were suppressed 10 to 30-fold compared to those of DBA/1 mice not injected with IL-12. CII-specific antibodies of the IgG2b subclass were also downregulated (3- to 10-fold), whereas those of the IgG2a subclass remained high. These results demonstrate that the mitigation of CII-induced arthritis in DBA/1 mice immunized CII in CFA by high dose IL-12 treatment is associated with an unique modification of the humoral immune reaction to CII.

DISCUSSION

Pathogenesis of Arthritic Diseases and Th1 Type Immune Reactivity

Arthritic diseases appear to be associated with immune responses of the Th1 type. Several groups succeeded in isolating pathogen-specific Th1 clones from patients with reactive arthritis.[70-72] In a mouse model of experimental Lyme disease,[73] infection of BALB/c mice with Borrelia burgdorferi resulted in IL-4 production and correlated with resistance. In contrast, infected C3H/HeJ mice produced IFN-γ and developed severe arthritis. Most of the T cell lines established from peripheral blood or synovial tissue of patients with rheumatoid arthritis also produced Th0/Th1 cytokines.[74-76] A Th1 type cytokine pattern was detected by in situ hybridization in the synovial tissue from rheumatoid arthritis patients.[77] In the CII-induced arthritis model, spleen cells from CII in CFA-immunized DBA/1 mice produce IFN-γ upon restimulation with CII ex vivo[60] and in CII in CFA-immunized animals treated with a nonmitogenic anti-CD3 mAb the synthesis of IL-2 and IFN-γ but not of IL-4 was markedly decreased.[78] This was associated with a delayed onset and reduced severity of arthritis.

In agreement with the assumption that arthritic diseases are linked to Th1 type autoreactivity, administration of IL-12 promoted severe joint disease in DBA/1 mice by enhancing the cellular and humoral response to CII. Immunization of DBA/1 mice with CII in IFA results in an intermediate immune response with a low incidence of arthritis (0-20%). Mainly CII-specific antibodies of the IgG1 subclass were detected in the sera of these animals. The CII-specific synthesis of IFN-γ by ex vivo activated spleen cells was rather low. In animals additionally treated with IL-12, the incidence of arthritis increased to 80-100%. Furthermore, the CII-specific synthesis of IFN-γ by splenic CD4-positive T cells and the titers of CII-specific antibodies in their sera were much higher. The most pronounced effect of IL-12 treatment was observed for antibodies of the IgG2a subclass that were upregulated 10 to 100-fold. Thus, IL-12 primarily enhanced the synthesis of antibodies of the complement-fixing subclass(es), that are important in the pathogenesis of CII-induced arthritis.[57,62,63,79] Human CII-reactive antibodies secreted by mononuclear cells isolated from inflamed synovial tissue of patients suffering from rheumatoid arthritis[65] may play a similar role in the pathogenesis of this disease.

The Relative Importance of Autoantigen-Specific Antibodies and Th1 Cells in CII-Induced Arthritis and Other Autoimmune Diseases

There is convincing evidence that autoantibodies are directly causing certain autoimmune diseases such as myasthenia gravis or hyperthyroidism as discussed in the Introduction section. However, their role in the pathogenesis of arthritis, diabetes and EAE is less clear. Activation of B cells/production of autoantibodies is frequently observed but this may be an epiphenomenon not causatively linked to the development of the respective disease. In organ-specific autoimmune diseases such as diabetes and EAE, "cellular" Th1 type responses are considered to be pathogenic and "humoral" Th2 type responses to be protective.[6,31] The evidence for pathogenic Th1 cells in the development of arthritic diseases has already been discussed. However, at least in the CII-induced arthritis model the pathogenesis of joint disease appears to depend on a "cellular" and "humoral" Th1 response. Under two experimental conditions, namely in DBA/1 mice immunized with CII in IFA + IL-12 + anti-IFN-γ mAb[66,67] and in mice with a C57BL background immunized with CII in IFA + IL-12,[68] we observed a very strong CII-specific Th1 cell response but only occasionally symptoms of joint disease. Surprisingly, the synthesis of IFN-γ by spleen cells from CII in IFA-immunized DBA/1 mice treated with IL-12 plus anti-IFN-γ mAb was even higher than that of spleen cells from animals additionally treated with IL-12, which is in agreement with results showing enhanced Th1 cell development in IFN-γ receptor-deficient mice infected with parasites[80] or virus.[45] Under these conditions, there was virtually no enhancing effect of IL-12 on the synthesis of CII-specific antibodies and the serum titers of the complement-fixing antibody subclasses IgG2a and IgG2b remained low. Thus, in CII in IFA-immunized DBA/1 mice treated with IL-12 + anti IFN-γ mAb, IL-12-induced IFN-γ appears to be primarily required for the enhancement of the "humoral" but not the "cellular" response to CII and the mere presence of Th1 cells is not sufficient to induce arthritis.

Although diabetes (IDDM) and EAE/multiple sclerosis are considered to be induced by autoreactive T cells and transfer of immune serum alone does not induce disease, autoantibodies are produced in large amounts during the course of these diseases. In animals[81] as well as in IDDM patients[82] autoantibodies against the islets of Langerhans are detectable very early before the onset of diabetes. The antibody response to human myelin basic protein, a candidate antigen in multiple sclerosis, is directed to the same (main) epitope as the T cell response.[83] Direct evidence for a role of autoantibodies in the pathogenesis of EAE comes from a rat model of EAE. Demyelinating antibodies determined the size and structure of brain lesions when transferred in combination with autoreactive T cells.[84] Furthermore, in a nonhuman primate model of EAE, immunization with myelin basic protein resulted in inflammation with little demyelination. Fully demyelinated lesions were observed upon injection of autoantibodies (to a second antigen) in combination with encephalitogenic T cells, suggesting that the T cells may contribute to EAE pathology by disrupting of the blood-brain barrier enabling antibodies to enter the central nervous system.[85] One should also take into consideration that not the overall amount of autoantibodies generated, but rather the IgG subclass, may determine the pathogenic potential of autoantibodies in these diseases. Complement-fixing antibody subclasses may be more effective in promoting disease than noncomplement-fixing antibodies. Indeed, in a model of antibody-mediated demyelinating EAE, treatment with a soluble recombinant complement receptor type 1 reduced clinical severity, inhibited inflammation, and almost completely blocked demyelination.[86] Using a mAb that blocks the complement cascade at C5, thereby preventing the generation of the potent chemoattractant C5a, Wang et al[87] could prevent CII-induced arthritis in DBA/1 mice immunized with CII in CFA. Moreover, anti-C5 antibody treatment initiated at the time of established disease ameliorated clinical and histological symptoms in the affected joints and prevented the recruitment of additional joints. Thus, complement-fixing autoantibodies in combination with autoreactive T cells may contribute to the immunopathogenesis of autoimmune diseases such as arthritis, diabetes, and encephalomyelitis in humans and animals.

To definitely explore the pathologic relevance of autoantigen-specific B cells as producers of autoantibodies or autoantigen-presenting cells in the development of these diseases, studies with B cell knockout mice in the several models of EAE, arthritis, and diabetes are required.

Regulation of Humoral Immunity/Antigen-Specific IgG2a Production by IL-12/IFN-γ

The involvement of IFN-γ in the regulation of "cellular" and "humoral" autoimmune responses has been extensively discussed above. In the CII-induced arthritis model, IFN-γ appears to be primarily required for the IL-12-induced enhancement of the "humoral" but not the "cellular" immune reaction to CII. The upregulation of the antibody/IgG2a synthesis by IL-12 via the stimulation of IFN-γ synthesis probably reflects IFN-γ's activity as a mediator promoting IgG2a synthesis.[18,19,24,25] In addition to the induction of IFN-γ, IL-12 may also have some direct effects on B cells as demonstrated for human B cells.[88] Compared to many other cytokines, IL-12 in combination with IL-2 was superior in supporting proliferation and IgG synthesis of purified human B cells stimulated in vitro with *Staphyloccocus aureus Cowan*. Thus, the IL-12-induced synthesis of IFN-γ, direct effects of IL-12, and the higher frequency of antigen-specific Th1 cells in IL-12-treated animals immunized with protein antigens may all contribute to the augmentation of antibody synthesis observed in immunized animals treated with IL-12.

However, the effects of IL-12 on antibody responses are heterogenous. Initially, mainly inhibitory influences of IL-12 on antibody synthesis were described. IL-12 downregulated the production of human IgE in an in vitro model of polyclonal IgE synthesis.[89] Administration of IL-12 suppressed serum IgG and IgE levels in a murine in vivo model of polyclonal B cell activation by treatment with anti-IgD antibodies.[90] Shortly after immunization of mice with a hapten-protein conjugate plus IL-12 treatment, the synthesis of IL-4 and of hapten-specific IgG1 were markedly suppressed as compared to non-IL-12-treated control animals.[91] The synthesis of IFN-γ was strongly, and that of IgG2a antibodies slightly, augmented in IL-12-treated mice during the primary response. Later on, this group reported that IL-12 treatment (in contrast to the effects observed during the primary response) enhanced the Th1 (IFN-γ, IgG2a) as well as the Th2 (IL-4, IgG1) recall response to a hapten-protein conjugate.[92] Similar results were obtained by Buchanan et al.[93] Compared to PBS-treated control animals, administration of IL-12 greatly elevated antigen-specific IgG2a and reduced IgG1 at day seven after immunization with a protein antigen. On day 28 after immunization, the serum levels of specific IgG1 and IgG2a antibodies were both increased in IL-12-treated mice. Thus, some of the variations regarding the influence of IL-12 on antibody synthesis can be explained by the time dependency of its effects. The results of our group describing the adjuvant effect of IL-12 for the synthesis of the Th1 type IgG2a/2b/3 antibodies[66,94] and the failure of IL-12 to permanently suppress antigen-specific IgE/Th2 responses[95] are in full agreement with these reports.

Under certain circumstances, IL-12 appears to be able to induce a more long-lasting suppression of "humoral" immunity. In mice transgenic for the hepatitis B *e* antigen, injection of a peptide (to which no T cell tolerance exists) induced a hepatitis B *e* antigen-specific antibody response of the IgG1 subclass and barely detectable IgG2a and IgG2b antibodies.[96] Additional treatment with IL-12 (1 μg/daily) resulted in the production of some IgG2a and IgG2b antibodies but suppressed the IgG1 response almost completely. This suppression was still observed even 7 weeks after priming and was accompanied by a profound inhibition of IL-4 synthesis.[96] The IgG subclass pattern induced by IL-12 in this transgenic model is very similar to the one found in DBA/1 mice immunized with CII in CFA and treated for at least 2 weeks with high doses (1 μg/daily) of IL-12. Under these conditions, IL-12 treatment resulted in a long-lasting suppression of CII-specific IgG1 antibodies, slightly reduced IgG2b serum levels, and exerted no effect on IgG2a levels, compared to DBA/1 mice immunized with CII in CFA without IL-12.[69] The most likely explanation is that relatively high levels of IFN-γ are present in these mice that almost completely inhibit the switching of B cells to IgG1. That IFN-γ can downregulate IgG1 synthesis has been described in a different in vivo model[24] and several in vitro models.[18,19,21] In those cases where injections of IL-12 inhibited IgG1 early during the priming response but not at later stages, the inhibition of IgG1 isotype switching in B cells may have been incomplete, allowing their expansion and synthesis of IgG1 antibodies at later stages. Indeed, neutralization of IFN-γ relieved the early inhibition of antigen-specific IgG1 production in IL-12-treated mice in two,[90,93] but not a third,[91] study. Clearly, experiments with IFN-γ receptor knockout mice are necessary to determine whether or not IL-12 by itself exerts some inhibitory or stimulatory effect on antibody synthesis.

Frequently, IL-12 is considered to be a cytokine promoting "cellular" immunity.[17] The results discussed above suggest rather that IL-12 promotes "cellular" and "humoral" Th1 type immune reactivity. Interestingly, in contrast to DBA/1 mice, IL-12 treatment failed to upregulate the CII-specific antibody synthesis and to promote arthritis in CII in IFA-immunized mice with C57BL background in a highly reproducible manner.[68] Since the development of IFN-γ-producing T cells was equivalent in IL-12-injected DBA/1 and C57BL mice, these results suggest that either a cofactor required for the IL-12-induced upregulation of antibody synthesis is not generated during immunization of C57BL/6 mice with CII in IFA, or, alternatively, that the CII-specific Th1 cells developing in DBA/1 and C57BL mice are heterogenous regarding their helper function for B cells. It is important to note that the failure of IL-12 to enhance the CII-specific antibody/IgG2a response in mice with C56BL background was specific for CII as antigen. Several strains of mice including C57BL/6 or C57BL10.Q produce large amounts of specific IgG1 and very little IgG2a when immunized with keyhole limpet hemocyanin adsorbed to aluminum hydroxide or emulsified with IFA. Administration of IL-12 caused a tremendous (> 100-fold) increase in the synthesis of IgG2a specific for keyhole limpet hemocyanin.[97] Thus, although the inability of IL-12 to upregulate the synthesis of CII-specific IgG2a is due to the influence of so far unknown non-MHC background genes of the C57BL genome, this phenomenon is also antigen-specific. Much effort is necessary to identify the underlying mechanisms that distinguish the helper function of different Th1 cells and that discriminate between "humoral" and "cellular" immunity, in particular since this may have consequences for the understanding of the pathogenesis of at least some autoimmune diseases.

Downregulation of Autoimmune Responses by IL-12 / IFN-γ

Unexpectedly, systemic administration of high doses of IL-12 for at least two weeks during the priming immunization with CII in CFA reduced incidence and severity of CII-induced arthritis in DBA/1 mice.[69] This was associated with a suppression of the humoral anti-CII response. The IgG subclass most significantly reduced was (CII-specific) IgG1, followed by IgG2b, and no effect on IgG2a. This IgG subclass pattern of CII-specific antibodies is clearly different from the one observed in DBA/1 mice immunized with CII in IFA, suggesting that different mechanisms account for the mitigation/absence of joint disease under these two conditions. Thus, IL-12 can both suppress and enhance autoimmune phenomena in CII-induced arthritis in DBA/1 mice, depending on the adjuvant.

IFN-γ can also promote or mitigate CII-induced arthritis. Mice locally treated with IFN-γ at day 16 after the priming immunization showed a higher frequency and severity of arthritis, compared to control animals.[98] Administration of low doses of IFN-γ at the time of immunization accelerated the onset and increased the incidence of disease in the CII-induced arthritis model.[99] Moreover, neutralization of IFN-γ during the first 4 weeks after immunization of DBA/1 mice with CII in CFA almost completely prevented arthritis.[100] Only very few infiltrating cells could be detected in the joints of anti-IFN-γ-treated animals and the titers of anti-CII antibodies in the sera were slightly reduced. In contrast, late treatment with a neutralizing anti-IFN-γ mAb, during the effector phase of the disease (week 5-8 after priming), exacerbated the severity of CII-induced arthritis.[100] Similar results were obtained in a second study where (partial) neutralization of IFN-γ increased the number of affected limbs.[101] Furthermore, continuous treatment of CII in CFA-immunized DBA/1 mice with high doses of IFN-γ started at day 20 after priming delayed the onset and downregulated the severity of joint disease.[102] These studies indicate that IFN-γ (like IL-12) can either aggravate or mitigate CII-induced arthritis, and it is tempting to speculate that the effects of IL-12 might be at least partially mediated via IFN-γ. The opposing effects of IL-12/IFN-γ may be explained by the dose and/or the time of administration. IL-12/IFN-γ appears to be obligatory early in the development of the arthritogenic CII-specific immune response induced in DBA/1 mice by immunization with either CII in CFA or CII in IFA + IL-12.[67,100] During later stages of the disease, IFN-γ[100-102] exerts a mitigating influence. This may indicate the development of some regulatory IFN-γ-producing T cells, similar to those described recently in NOD mice.[103] Alternatively, IL-12/IFN-γ may promote or suppress joint disease at low or high doses, respectively. The different effect of IL-12 on the course of CII-induced arthritis elicited by immunization with CII in IFA versus CII in CFA could be explained by higher levels of IFN-γ in the CII in CFA-immunized mice, due to the prolonged treatment with IL-12 and the immunization with two antigens (CII and CFA-mycobacteria) resulting in higher levels of IFN-γ. Indeed, the sera of IL-12-treated DBA/1 mice immunized with CII in CFA contain large amounts of IFN-γ (unpublished observation). The failure of late IL-12 treatment to suppress the severity of joint disease[69] may be explained by lower IFN-γ levels, as compared to early treatment, since the booster immunization is performed with CII in saline lacking mycobacteria. Experiments are currently under way investigating whether therapeutic injections of high doses of IL-12 will mitigate joint disease when the booster immunization is performed with CII in CFA and whether this correlates with higher serum levels of IFN-γ.

Additional properties of IL-12/IFN-γ that could conceivably contribute to their anti-arthritic effects are the induction of glucocorticoid synthesis by IL-12 in vivo[104] and the inhibition of monocyte/ macrophage recruitment into the synovium by high doses of systemically applicated IFN-γ due to a downregulation of receptors for the chemotactic complement component C5a.[105]

Different Patterns of Humoral CII-Specific Immune Responses and Their Relationship to the Development of CII-Induced Arthritis

The importance of CII-specific antibodies, especially of the complement-fixing IgG2a and IgG2b subclasses, in the pathogenesis of CII-induced arthritis has been extensively discussed. Treating CII-immunized mice with IL-12 under various conditions enabled us to discriminate between three subtypes of (humoral) CII-specific immunity, which differ regarding the magnitude and the IgG subclass pattern of the CII-reactive antibody response. Since only one subtype is associated with severe destructive arthritis, the precise characterization of CII-specific IgG serum levels and the determination of their subclasses may allow prognosis of whether or not the animals will develop arthritis (summarized in Table 2.1).

Conditions associated with the development of severe joint disease are characterized by high serum levels of the three IgG subclasses G1, G2a and G2b. This type of CII-specific antibody response will be referred to as "high, mixed" subtype. It is found in DBA/1 mice immunized with CII in CFA or CII in IFA + IL-12. However, a few DBA/1 mice display this subtype of antibody response to CII, yet do not develop arthritis. Likewise, certain strains of Biozzi mice are resistant to CII-induced arthritis despite mounting a "high, mixed" type antibody response.[106] This indicates that other factors (in combination with antibodies) are essential in the pathogenesis of arthritis, or that regulatory T cells can prevent joint disease even in the presence of a strong anti-CII immune response.

The second type of humoral CII-specific immunity is characterized by about 3 to 10-fold lower serum levels of CII-specific antibodies. IgG1 is the dominant subclass, in combination with low titers of IgG2a and IgG2b antibodies. Animals showing this "intermediate" subtype of humoral anti-CII response do not (or only occasionally) develop arthritis. This type is observed in DBA/1 mice immunized with CII in IFA, in DBA/1 mice immunized with CII in IFA and treated with IL-12 in combination with a neutralizing anti-IFN-γ mAb,[66,67] and in IL-12-treated mice with C57BL background injected with CII in IFA.[68] Likewise, DBA/1 mice suppressed for CII-induced arthritis by pretreatment with CII displayed an about 10-fold lower CII-specific antibody response which was mainly restricted to the IgG1 subclass.[107] In another system, the amelioration of arthritis observed in DBA/1 mice rendered hyporesponsive for Th1 but not Th2 cells was not associated with a decrease in the overall CII-specific antibody production.[78] However, it is likely that the antibody isotype pattern was altered, resulting in high IgG1/IgE and low IgG2a synthesis. Such a shift away from the synthesis of complement-fixing antibody isotypes could very well be responsible for the reduced severity of joint disease observed in these animals.

Somewhat unexpected was the finding that CII-induced arthritis occurs with lower incidence and severity in DBA/1 immunized with CII in CFA + IL-12, despite the development of a rather polarized "high, Th1" type response.[69] Under these conditions, the overall CII-specific antibody synthesis was slightly reduced compared to DBA/1 mice immunized with CII in CFA or CII in IFA + IL-12. Antibodies of the IgG2a subclass remained high, whereas antibodies of

Different subtypes of humoral responses to type II collagen - correlations with the development of arthritis

Subtype	Ig isotype pattern: IgG1	IgG2a	IgG2b	Arthritis
"intermediate"	++	+	+	+/-
(DBA/1 mice: CII in IFA or CII in IFA + IL-12 + anti-IFN-γ) (C57BL mice: CII in CFA or CII in IFA -/+ IL-12)				
"high, mixed"	+++	+++	+++	+++
(DBA/1 mice: CII in CFA or CII in IFA + IL-12)				
"high, Th1"	+/-	+++	++	+
(DBA/1 mice: CII in CFA + IL-12)				

Table 2.1.

the IgG1 subclass were strongly, and that of the IgG2b subclass modestly, inhibited, clearly distinguishing this type of humoral anti-CII response from those described before. Interestingly, there are a few reports in the literature describing strong anti-CII Th1/IgG2a responses in the absence of joint disease.[106,108,109] In two strains of Biozzi mice selected for high antibody responsiveness to sheep erythrocytes, cellular and humoral CII-specific immune reactions could be elicited but only the animals of one strain suffered from joint disease.[108] In the susceptible Biozzi strain, IFN-γ as well as IL-5-producing T cells could be detected, whereas T cells secreting IFN-γ were high but those secreting IL-5 much lower in the spleens from mice of the resistant Biozzi strain. Comparable (high) amounts of CII-specific IgG1 and IgG2a were present, the sera of mice of both strains. These results also support the concept that rheumatoid-like arthritis in mice is associated rather with a "mixed" than a polarized "Th1" type of immune reactivity to CII. Likewise, CII in CFA-immunized transgenic SWR mice, expressing the TCR β chain of a CII-reactive T cell clone, did not develop arthritis despite mounting a high (IgG2a) antibody response to CII, comparable or even higher than the one observed in DBA/1 mice.[109] The resistance of (transgenic) SWR mice to the induction of CII-induced arthritis is probably related to their complement component C5 deficiency[63,110] as well as the influence of other genes important for the induction and maintenance of tolerance to CII.[110] In contrast to (DBA/1xC57BL/6)F1 mice, F1 crosses between SWR and DBA/1 are relatively resistant to the induction of CII-induced arthritis,[110] suggesting that different mechanisms may account for the resistance of mice with SWR or C56BL background to the induction of CII-induced arthritis.

Taken together, the results discussed demonstrate the complexity of the pathogenesis of the autoimmune response to CII leading to joint disease. Although neither polarized "Th2 type" nor polarized "Th1 type" immune reactivity to CII appears to result in arthritis, many of the components comprising and regulating the "high, mixed" subtype of the CII-specific immune response associated with arthritis remain to be identified.

ACKNOWLEDGMENTS

The author would like to thank all collaborators, especially Henry Hess and Jacek Szeliga who performed most of the experiments described herein. I would also like to express my gratitude to Maurice K. Gately and Frank J. Podlaski, Hoffmann-LaRoche Inc., Nutley, USA for the preparation and generous gift of mouse recombinant IL-12 and anti-IL-12 reagents.

This work was supported by the Deutsche Forschungsgemeinschaft, Sonderforschungsbereich 311. Jacek Szeliga was supported by a fellowship from Deutscher Akademischer Austauschdienst.

REFERENCES

1. Wraith DC, McDevitt HO, Steinman L et al. T cell recognition as the target for immune intervention in autoimmune disease. Cell 1989; 57:709-715.
2. Bach JF. The multifactorial etiology of autoimmune diseases. Clin Exp Rheumatol 1994; 12 Suppl 11:S5-8.
3. Steinman L. Escape from "horror autotoxicus": pathogenesis and treatment of autoimmune disease. Cell 1995; 80:7-10.
4. Miller RD, Calkins CE. Development of self-tolerance in normal mice. Appearance of suppressor cells that maintain adult self-tolerance follows the neonatal autoantibody response. J Immunol 1988; 141:2206-2210.
5. Naparstek Y, Plotz PH. The role of autoantibodies in autoimmune disease. Annu Rev Immunol 1993; 11:79-104.
6. Liblau RS, Singer SM, McDevitt HO. Th1 and Th2 CD4+ T cells in the pathogenesis of organ-specific autoimmune diseases. Immunol Today 1995; 16:34-38.
7. Pfeiffer C, Stein J, Southwood S et al. Altered peptide ligands can control CD4 T lymphocyte differentiation in vivo. J Exp Med 1995; 181:1569-1574.
8. Constant S, Pfeiffer C, Woodard A et al. Extent of T cell receptor ligation can determine the functional differentiation of naive CD4+ T cells. J Exp Med 1995; 182:1591-1596.
9. Hosken NA, Shibuya K, Heath AW et al. The effect of antigen dose on CD4+ T helper cell phenotype development in a T cell receptor-alpha beta-transgenic model. J Exp Med 1995; 182:1579-1584.
10. Hsieh CS, Macatonia SE, Tripp CS et al. Development of TH1 CD4+ T cells through IL-12 produced by Listeria-induced macrophages [see comments]. Science 1993; 260:547-549.
11. Schmitt E, Hoehn P, Huels C et al. T helper type 1 development of naive CD4+ T cells requires the coordinate action of interleukin-12 and interferon-gamma and is inhibited by transforming growth factor-beta. Eur J Immunol 1994; 24:793-798.
12. Mosmann TR, Cherwinski H, Bond MW et al. Two types of murine helper T cell clone. I. Definition according to profiles of lymphokine activities and secreted proteins. J Immunol 1986; 136:2348-2357.
13. Bottomly K. A functional dichotomy in CD4+ T lymphocytes [see comments]. Immunol Today 1988; 9:268-274.
14. Cher DJ, Mosmann TR. Two types of murine helper T cell clone. II. Delayed-type hypersensitivity is mediated by TH1 clones. J Immunol 1987; 138:3688-3694.
15. Tsicopoulos A, Hamid Q, Varney V et al. Preferential messenger RNA expression of Th1-type cells (IFN-gamma+, IL-2+) in classical delayed-type (tuberculin) hypersensitivity reactions in human skin. J Immunol 1992; 148:2058-2061.

16. Stout RD, Bottomly K. Antigen-specific activation of effector macrophages by IFN-gamma producing (TH1) T cell clones. Failure of IL-4-producing (TH2) T cell clones to activate effector function in macrophages. J Immunol 1989; 142:760-765.
17. Scott P. IL-12: initiation cytokine for cell-mediated immunity [comment]. Science 1993; 260:496-497.
18. Snapper CM, Paul WE. Interferon-gamma and B cell stimulatory factor-1 reciprocally regulate Ig isotype production. Science 1987; 236:944-947.
19. Stevens TL, Bossie A, Sanders VM et al. Regulation of antibody isotype secretion by subsets of antigen-specific helper T cells. Nature 1988; 334:255-258.
20. Parish CR. Immune response to chemically modified flagellin. II. Evidence for a fundamental relationship between humoral and cell-mediated immunity. J Exp Med 1971; 134:21
21. Murray JS, Madri J, Pasqualini T, Bottomly K. Functional CD4 T cell subset interplay in an intact immune system. J Immunol 1993; 150:4270-4276.
22. Rizzo LV, DeKruyff RH, Umetsu DT et al. Regulation of the interaction between Th1 and Th2 cell clones to provide help for antibody production in vivo. Eur J Immunol 1995; 25:708-716.
23. Mahon BP, Katrak K, Nomoto A et al. Poliovirus-specific CD4+ Th1 clones with both cytotoxic and helper activity mediate protective humoral immunity against a lethal poliovirus infection in transgenic mice expressing the human poliovirus receptor. J Exp Med 1995; 181:1285-1292.
24. Finkelman FD, Katona IM, Mosmann TR, Coffman RL. IFN-gamma regulates the isotypes of Ig secreted during in vivo humoral immune responses. J Immunol 1988; 140:1022-1027.
25. Huang S, Hendriks W, Althage A et al. Immune response in mice that lack the interferon-gamma receptor [see comments]. Science 1993; 259:1742-1745.
26. Snapper CM, McIntyre TM, Mandler R et al. Induction of IgG3 secretion by interferon gamma: a model for T cell-independent class switching in response to T cell-independent type 2 antigens. J Exp Med 1992; 175:1367-1371.
27. Swain SL, Weinberg AD, English M et al. IL-4 directs the development of Th2-like helper effectors. J Immunol 1990; 145:3796-3806.
28. Seder RA, Paul WE, Davis MM et al. The presence of interleukin 4 during in vitro priming determines the lymphokine-producing potential of CD4+ T cells from T cell receptor transgenic mice. J Exp Med 1992; 176:1091-1098.
29. Trinchieri G. Interleukin-12: A cytokine produced by antigen-presenting cells with immunoregulatory functions in the generation of T-helper cells type 1 and cytotoxic lymphocytes. Blood 1994; 84:4008-4027.
30. Germann T, Rüde E. Interleukin 12. Int Arch Allergy Immunol 1995; 108:103-112.

31. Trembleau S, Germann T, Gately MK, Adorini L. The role of IL-12 in the induction of organ-specific autoimmune diseases. Immunol Today 1995; 16:383-386.
32. Marcelletti JF, Ohara J, Katz DH. Collagen-induced arthritis in mice. Relationship of collagen-specific and total IgE synthesis to disease. J Immunol 1991; 147:4185-4191.
33. Boissier MC, Bessis N, Caput D et al. Gene therapy using IL-4 and IL-13 gene-transfected cells in murine collagen-induced arthritis. Arthritis and Rheumatism 1994; 37:215 (Abstract)
34. Allen JB, Wong HL, Costa GL et al. Suppression of monocyte function and differential regulation of IL-1 and IL-1ra by IL-4 contribute to resolution of experimental arthritis. J Immunol 1993; 151:4344-4351.
35. D'Andrea A, Aste Amezaga M, Valiante NM et al. Interleukin 10 (IL-10) inhibits human lymphocyte interferon gamma-production by suppressing natural killer cell stimulatory factor/IL-12 synthesis in accessory cells. J Exp Med 1993; 178:1041-1048.
36. de Waal Malefyt R, Figdor CG et al. Effects of IL-13 on phenotype, cytokine production, and cytotoxic function of human monocytes. Comparison with IL-4 and modulation by IFN-gamma or IL-10. J Immunol 1993; 151:6370-6381.
37. Katsikis PD, Chu CQ, Brennan FM et al. Immunoregulatory role of interleukin 10 in rheumatoid arthritis. J Exp Med 1994; 179:1517-1527.
38. Chomarat P, Vannier E, Dechanet J et al. Balance of IL-1 receptor antagonist/IL-1 beta in rheumatoid synovium and its regulation by IL-4 and IL-10. J Immunol 1995; 154:1432-1439.
39. Trinchieri G, Perussia B. Immune interferon: a pleiotropic lymphokine with multiple effects. Immunol Today 1985; 6:131-136.
40. Dalton DK, Pitts-Meek S et al. Multiple defects of immune cell function in mice with disrupted interferon-gamma genes [see comments]. Science 1993; 259:1739-1742.
41. Coutelier JP, van der Logt JT, Heessen FW. IgG2a restriction of murine antibodies elicited by viral infections. J Exp Med 1987; 165:64-69.
42. Coutelier JP, Coulie PG, Wauters P et al. In vivo polyclonal B-lymphocyte activation elicited by murine viruses. J Virol 1990; 64:5383-5388.
43. Scherle PA, Palladino G, Gerhard W. Mice can recover from pulmonary influenza virus infection in the absence of class I-restricted cytotoxic T cells. J Immunol 1992; 148:212-217.
44. Graham MB, Braciale VL, Braciale TJ. Influenza virus-specific CD4+ T helper type 2 T lymphocytes do not promote recovery from experimental virus infection. J Exp Med 1994; 180:1273-1282.
45. Schijns VE, Haagmans BL, Rijke EO et al. IFN-gamma receptor-deficient mice generate antiviral Th1-characteristic cytokine profiles but altered antibody responses. J Immunol 1994; 153:2029-2037.
46. Graham MB, Dalton DK. Response to influenza infection in mice with a targeted disruption in the interferon-g gene. J Exp Med 1993; 178:1725-1732.

47. Finkelman FD, Svetic A, Gresser I et al. Regulation by interferon alpha of immunoglobulin isotype selection and lymphokine production in mice. J Exp Med 1991; 174:1179-1188.
48. Gu D, Wogensen L, Calcutt NA et al. Myasthenia gravis-like syndrome induced by expression of interferon gamma in the neuromuscular junction. J Exp Med 1995; 181:547-557.
49. Campbell IL, Kay TW, Oxbrow L et al. Essential role for interferon-gamma and interleukin-6 in autoimmune insulin-dependent diabetes in NOD/Wehi mice. J Clin Invest 1991; 87:739-742.
50. Debray Sachs M, Carnaud C, Boitard C et al. Prevention of diabetes in NOD mice treated with antibody to murine IFN gamma. J Autoimmun 1991; 4:237-248.
51. Sarvetnick N, Shizuru J, Liggitt D et al. Loss of pancreatic islet tolerance induced by beta-cell expression of interferon-gamma. Nature 1990; 346:844-847.
52. Billiau A, Heremans H, Vandekerckhove F et al. Enhancement of experimental allergic encephalomyelitis in mice by antibodies against IFN-gamma. J Immunol 1988; 140:1506-1510.
53. Ferber IA, Brocke S, Taylor-Edwards C et al. Mice with a disrupted IFN-γ gene are susceptible to the induction of experimental autoimmune encephalomyelitis (EAE). J Immunol 1996; 156:5-7.
54. Caspi RR, Chan CC, Grubbs BG et al. Endogenous systemic IFN-gamma has a protective role against ocular autoimmunity in mice. J Immunol 1994; 152:890-899.
55. Trentham DE, Townes AS, Kang AH. Autoimmunity to type II collagen: An experimental model of arthritis. J Exp Med 1977; 146:857-875.
56. Courtenay JS, Dallman MJ, Dayan AD et al. Immunisation against heterologous type II collagen induces arthritis in mice. Nature 1980; 283:666-668.
57. Wooley PH, Luthra HS, Stuart JM, David CS. Type II collagen-induced arthritis in mice I. Major histocompatibility complex (I region) linkage and antibody correlates. J Exp Med 1981; 154:688-700.
58. Seki N, Sudo Y, Mizuhara H et al. Type II collagen-induced murine arthritis: induction of arthritis depends on antigen-presenting cell function as well as susceptibility of host to an anticollagen immune response. J Immunol 1992; 148:3093-3099.
59. Ranges GE, Sriram S, Cooper SM. Prevention of type II collagen-induced arthritis by in vivo treatment with anti-L3T4. J Exp Med 1985; 162:1105-1110.
60. Myers LK, Rosloniec EF, Seyer JM et al. A synthetic peptide analogue of a determinant of type II collagen prevents the onset of collagen-induced arthritis. J Immunol 1993; 150:4652-4658.
61. Jansson L, Holmdahl R. Genes on the X chromosome affect development of collagen-induced arthritis in mice. Clin Exp Immunol 1993; 94:459-465.

62. Durie FH, Fava RA, Foy TM et al. Prevention of collagen-induced arthritis with an antibody to gp39, the ligand for CD40. Science 1993; 261:1328-1330.
63. Watson WC, Townes AS. Genetic susceptibility to murine collagen II autoimmune arthritis. Proposed relationship to the IgG2 autoantibody subclass response, complement C5, major histocompatibility complex (MHC) and non-MHC loci. J Exp Med 1985; 162:1878-1891.
64. Trentham DE. Collagen arthritis as a relevant model for rheumatoid arthritis. Arthritis Rheum 1982; 25:911
65. Tarkowski A, Klareskog L, Carlsten H et al. Secretion of antibodies to types I and II collagen by synovial tissue cells in patients with rheumatoid arthritis. Arthritis Rheum 1989; 32:1087-1092.
66. Germann T, Szeliga J, Hess H et al. Administration of IL-12 in combination with type II collagen induces severe arthritis in DBA/1 Mice. Proc Natl Acad Sci U S A 1995; 92:4823-4827.
67. Germann T, Hess H, Szeliga J et al. Characterization of the adjuvant effect of IL-12 and efficacy of IL-12 inhibitors in type II collagen-induced arthritis. Ann NY Acad Sci 1996; in press:
68. Szeliga J, Hess H, Rüde E et al. Interleukin-12 promotes cellular but not humoral type II collagen-specific Th1 type responses in C56BL/6 and B10.Q mice and fails to induce arthritis. Int Immunol 1996; 8:1221-1227.
69. Hess H, Gately MK, Rüde E et al. High doses of interleukin-12 inhibit the development of joint disease in DBA/1 mice immunized with type II collagen in complete Freund's adjuvant. Eur J Immunol 1996; 26:187-191.
70. Yssel H, Shanafelt MC, Soderberg C et al. Borrelia burgdorferi activates a T helper type 1-like T cell subset in Lyme arthritis. J Exp Med 1991; 174:593-601.
71. Schlaak J, Hermann E, Ringhoffer M et al. Predominance of Th1-type T cells in synovial fluid of patients with Yersinia-induced reactive arthritis. Eur J Immunol 1992; 22:2771-2776.
72. Lahesmaa R, Yssel H, Batsford S et al. Yersinia enterocolitica activates a T helper type 1-like T cell subset in reactive arthritis. J Immunol 1992; 148:3079-3085.
73. Matyniak JE, Reiner SL. T helper phenotype and genetic susceptibility in experimental Lyme disease. J Exp Med 1995; 181:1251-1254.
74. Miltenburg AM, van Laar JM, de Kuiper R et al. T cells cloned from human rheumatoid synovial membrane functionally represent the Th1 subset. Scand J Immunol 1992; 35:603-610.
75. Quayle AJ, Chomarat P, Miossec P et al. Rheumatoid inflammatory T-cell clones express mostly Th1 but also Th2 and mixed (Th0) cytokine patterns. Scand J Immunol 1993; 38:75-82.
76. Schulze-Koops H, Lipsky PE, Kavanaugh AF et al. Elevated Th1- or Th0-like cytokine mRNA in peripheral circulation of patients with rheumatoid arthritis. J Immunol 1995; 155:5029-5037.

77. Simon AK, Seipelt E, Sieper J. Divergent T-cell cytokine patterns in inflammatory arthritis. Proc Natl Acad Sci U S A 1994; 91:8562-8566.
78. Hughes C, Wolos JA, Giannini EH et al. Induction of T helper cell hyporesponsiveness in an experimental model of autoimmunity by using nonmitogenic anti-CD3 monoclonal antibody. J Immunol 1994; 153:3319-3325.
79. Seki N, Sudo Y, Yoshioka T et al. Type II collagen-induced murine arthritis. I. Induction and perpetuation of arthritis require synergy between humoral and cell-mediated immunity. J Immunol 1988; 140:1477-1484.
80. Swihart K, Fruth U, Messmer N et al. Mice from a genetically resistant background lacking the interferon gamma receptor are susceptible to infection with Leishmania major but mount a polarized T helper cell 1-type CD4+ T cell response. J Exp Med 1995; 181:961-971.
81. Baekkeskov S, Dyrberg T, Lernmark A. Autoantibodies to a 64-kilodalton islet cell protein precede the onset of spontaneous diabetes in the BB rat. Science 1984; 224:1348-1350.
82. Baekkeskov S, Landin M, Kristensen JK et al. Antibodies to a 64,000 Mr human islet cell antigen precede the clinical onset of insulin-dependent diabetes. J Clin Invest 1987; 79:926-934.
83. Warren KG, Catz I, Steinman L. Fine specificity of the antibody response to myelin basic protein in the central nervous system in multiple sclerosis: the minimal B-cell epitope and a model of its features. Proc Natl Acad Sci U S A 1995; 92:11061-11065.
84. Lassmann H, Brunner C, Bradl M et al. Experimental allergic encephalomyelitis: the balance between encephalitogenic T lymphocytes and demyelinating antibodies determines size and structure of demyelinated lesions. Acta Neuropathol Berl 1988; 75:566-576.
85. Genain CP, Nguyen M-H, Letvin NL et al. Antibody facilitation of multiple sclerosis-like lesions in nonhuman primate. J Clin Invest 1995; 96:2966-2974.
86. Piddlesden SJ, Storch MK, Hibbs M et al. Soluble recombinant complement receptor 1 inhibits inflammation and demyelination in antibody-mediated demyelinating experimental allergic encephalomyelitis. J Immunol 1994; 152:5477-5484.
87. Wang Y, Rollins SA, Madri JA et al. Anti-C5 monoclonal antibody therapy prevents collagen-induced arthritis and ameliorates established disease. Proc Natl Acad Sci U S A 1995; 92:8955-8959.
88. Jelinek DF, Braaten JK. Role of IL-12 in human B lymphocyte proliferation and differentiation. J Immunol 1995; 154:1606-1613.
89. Kiniwa M, Gately M, Gubler U et al. Recombinant interleukin-12 suppresses the synthesis of immunoglobulin E by interleukin-4 stimulated human lymphocytes. J Clin Invest 1992; 90:262-266.
90. Morris SC, Madden KB, Adamovicz JJ et al. Effects of IL-12 on in vivo cytokine gene expression and Ig isotype selection. J Immunol 1994; 152:1047-1056.

91. McKnight AJ, Zimmer GJ, Fogelman I et al. Effects of IL-12 on helper T cell-dependent immune responses in vivo. J Immunol 1994; 152:2172-2179.
92. Bliss J, Van Cleave V, Murray K et al. IL-12, as an adjuvant, promotes a T helper 1 cell, but does not suppress a T helper 2 cell recall response. J Immunol 1996; 156:887-894.
93. Buchanan JM, Vogel LA, Van Cleave VH et al. Interleukin 12 alters the isotype-restricted antibody response of mice to hen eggwhite lysozyme. Int Immunol 1995; 7:1519-1528
94. Germann T, Bongartz M, Dlugonska H et al. IL-12 profoundly upregulates the synthesis of antigen-specific, complement-fixing IgG2a, IgG2b and IgG3 antibody subclasses in vivo. Eur J Immunol 1995; 25:823-829.
95. Germann T, Guckes S, Bongartz M et al. Administration of interleukin-12 during ongoing immune responses fails to permanently suppress and can even enhance the synthesis of antigen-specific IgE. Int Immunol 1995; 7:1649-1657.
96. Milich DR, Wolf SF, Hughes JL et al. Interleukin 12 suppresses autoantibody production by reversing helper T-cell phenotype in hepatitis B e antigen transgenic mice. Proc Natl Acad Sci U S A 1995; 92:6847-6851.
97. Ramsdell F, Seaman MS, Miller RE et al. Differential ability of Th1 and Th2 T cells to express Fas ligand and to undergo activation-induced cell death. Int Immunol 1994; 6:1545-1553.
98. Mauritz NJ, Holmdahl R, Jonsson R et al. Treatment with gamma-interferon triggers the onset of collagen arthritis in mice. Arthritis Rheum 1988; 31:1297-1304.
99. Cooper SM, Sriram S, Ranges GE. Suppression of murine collagen-induced arthritis with monoclonal anti-Ia antibodies and augmentation with IFN-γ. J Immunol 1988; 141:1958-1962.
100. Boissier MC, Chiocchia G, Bessis N et al. Biphasic effect of interferon-gamma in murine collagen-induced arthritis. Eur J Immunol 1995; 25:1184-1190.
101. Williams RO, Williams DG, Feldmann M et al. Increased limb involvement in murine collagen-induced arthritis following treatment with anti-interferon-gamma. Clin Exp Immunol 1993; 92:323-327.
102. Nakajima H, Takamori H, Hiyama Y et al. The effect of treatment with interferon-gamma on type II collagen-induced arthritis. Clin Exp Immunol 1990; 81:441-445.
103. Akhtar I, Gold JP, Pan L et al. CD4+ beta islet cell-reactive T cell clones that suppress autoimmune diabetes in nonobese diabetic mice. J Exp Med 1995; 182:87-97.
104. Orange JS, Salazar-Mather TP, Opal SM et al. Mechanism of interleukin 12-mediated toxicities during experimental viral infections: role of tumor necrosis factor and glucocorticoids. J Exp Med 1995; 181:901-914.
105. Wahl SM, Allen JB, Ohura K et al. IFN-gamma inhibits inflammatory cell recruitment and the evolution of bacterial cell wall-induced arthritis. J Immunol 1991; 146:95-100.

106. Bouvet JP, Couderc J, Bouthillier Y et al. Collagen-induced arthritis in Biozzi mice. Joint involvement is not correlated with collagen II IgG2a autoantibodies nor restricted to only H-2q and H-2r. J Immunol 1989; 143:1537-1542.
107. Kresina TF, Finegan CK. Restricted expression of anti-type II collagen antibody isotypes in mice suppressed for collagen-induced arthritis. Ann Rheum Dis 1986; 45:60-66.
108. De Franco M, Gille Perramant MF et al. T helper subset involvement in two high antibody responder lines of mice (Biozzi mice): HI (susceptible) and HII (resistant) to collagen-induced arthritis. Eur J Immunol 1995; 25:132-136.
109. Mori L, Loetscher H, Kakimoto K et al. Expression of a transgenic T cell receptor beta chain enhances collagen-induced arthritis. J Exp Med 1992; 176:381-388.
110. Andersson M, Goldschmidt TJ, Michaelsson E et al. T-cell receptor V beta haplotype and complement component C5 play no significant role for the resistance to collagen-induced arthritis in the SWR mouse. Immunology 1991; 73:191-196.

CHAPTER 3

Th1/Th2 Balance in the Control of Drug-Induced Autoimmunity

Abdallah Badou, Lucette Pelletier and Philippe Druet

INTRODUCTION

CD4$^+$ T cells have been subdivided into Th1 and Th2 cells that differ by the profile of cytokines they produce, their function and their ability to cross-inhibit each other. Indeed, Th1 cells produce IL-2, IFN-γ and TNF-α while Th2 cells produce IL-4, IL-5, IL-6, IL-10 and IL-13.[1-4] IFN-γ inhibits the development of Th2 cells while IL-4 and IL-10 antagonize Th1 development (reviewed in refs. 1-5). The couple IL-12/IFN-γ sustains differentiation of Th precursors towards Th1 cells while IL-4 favors Th2 differentiation. Th1 cells are mainly involved in the development of cell-mediated immune responses while Th2 play a prominent role in B cell help for IgG1 and IgE isotype production and are involved in anaphylactic reactions. Th1 cells are responsible for organ-specific autoimmune diseases (reviewed in ref. 6) such as experimental autoimmune encephalomyelitis (EAE)[7] or experimental autoimmune uveoretinitis (EAU; Table 3.1).[8] EAE is induced by immunization with myelin basic protein in complete Freund's adjuvant and is characterized by transient paralysis while EAU is induced by immunization with S-retinal antigen and is characterized by an uveoretinitis. In the first autoimmune disease, Th2 cells which produce IL-4 and TGF-β are described as regulatory cells.[9-12] In this situation, Th2 cells may have a beneficial role because they produce cytokines with immunosuppressive properties and because they inhibit the development of pathogenic Th1 cells due to their production of IL-4. In experimental autoimmune encephalomyelitis, skewing the immune response towards a Th2 one was done by treating animals with IL-4[13], by targeting the presentation of the antigen by B cells[14] or by using pharmacological compounds (pentoxifylline or retinoic acid).[15, 16] It is noteworthy that Th1 cells seem to be responsible for systemic lupus erythematosus, a spontaneous antibody-mediated autoimmune disease that occurs in MRL

Immunointervention in Autoimmunity by Th1/Th2 Regulation,
edited by Luciano Adorini.

Table 3.1 Th1 and Th2 cells in immune responses

Th1	**Th2**
Organ-induced autoimmunity (EAE, EAU ...)	Autoimmunity associated with allogeneic reactions (HVG, GVH)
Murine systemic lupus erythematosus	Drug-induced autoimmunity
DTH reactions, induction of CTL differentiation	Anaphylactic reactions
Proinflammatory cytokines (IFN-γ, TNF-α)	Activation of mast cells, basophils (IL-4, IL-5) Anti-inflammatory cytokine : IL-10
B cell help (IgG2a, IgG3* fixing complement)	B cell help : (IgG1, IgE involved in anaphylactic reactions)

* = in mice
HVG : host versus graft reaction induced by injection of allogeneic cells into incompetent newborn mice
GVH : graft versus host reaction induced by the injection of parental cells C into (CxD) F1 hybrids. In some genetic combinations, T cells from C may polyclonally activate B lymphocytes of F1 (CxD) hybrids, leading to an autoimmune disease.

lpr/lpr[17] or NZB/NZW mice.[18] By contrast, Th2 cells clearly play a pathogenic role in autoimmunity associated with allogeneic reactions or in drug-induced autoimmunity, at least in BN rats, (reviewed in ref. 19; Table 3.1) . Several groups, including ours, were interested in the mechanisms responsible for the development of a Th2-mediated, drug-induced autoimmune disease in susceptible animals and for the resistance of other strains.

In this chapter, we will briefly present models of drug-induced immune disregulation. Then we will describe pathogenic autoreactive T cells implicated in the disease developed by susceptible animals and autoreactive T cells responsible for the immunosuppression found in resistant animals, and discuss the cytokines implicated. The direct effect of $HgCl_2$ on IL-4 gene transcription will be discussed. Finally, we will discuss the capacity of $HgCl_2$, used as an inducer of IL-4, to downgrade Th1-mediated autoimmune uveoretinitis .

MODELS

Drug-Induced Autoimmunity in Rats

Brown-Norway (BN) ($RT1^n$; RT1 = rat MHC) rats chronically treated with nontoxic doses of $HgCl_2$, gold salts or D-penicillamine display a systemic autoimmune disease marked by a polyclonal activation of B cells with a 20-fold increase in serum IgG1 concentration and a 1000-fold increase in serum IgE concentration.[20,21] These data suggest that Th2 cytokines are produced during the disease; by contrast, serum IgG2b concentration, likely a Th1-dependent isotype in rats, is reduced. Numerous autoantibodies are produced in the context of this activation, including IgG1 anti-laminin antibodies, which are found to be deposited in glomeruli and are associated with the development of a heavy proteinuria, at least in $HgCl_2$-treated rats.[22-25] The disease is milder in BN rats treated with gold salts[26] or D-penicillamine.[27] In addition, BN rats injected with $HgCl_2$ develop an early gut vasculitis which could be related to mast cell activation.[28] BN rats that survive spontaneously improve even if injections of chemicals are pursued. Rats in which $HgCl_2$ injections are stopped after two weeks become resistant to a subsequent challenge by $HgCl_2$ three months later.[29] MAXX ($RT1^n$) and DZB ($RT1^u$) rats which are genetically related to BN rats also display similar autoimmune manifestations upon $HgCl_2$ injection.[30,31] However, other strains such as LEW rats ($RT1^l$) are resistant.[24] In addition, LEW rats develop a nonantigen specific immunosuppression when injected with $HgCl_2$[32] and are protected against several organ-specific autoimmune diseases to which this strain is normally susceptible, such as experimental autoimmune encephalomyelitis (EAE)[33] or Heymann's nephritis.[34] $HgCl_2$-induced autoimmune disease or immunosuppression are under the dependence of T lymphocytes and the specificity of the autoreactive T cells implicated will be discussed below.

Analysis of cytokines in diseased rats shows an early (day 3), persistent (up to 30 days) and intense mRNA expression for IL-4 in the spleen and lymph nodes as demonstrated by quantitative PCR.[35] Other Th2 cytokines such as IL-6 and IL-10 are also upregulated but not IL-5 or IL-13 in diseased-BN rats as shown by semi-quantitative PCR.[36]

Drug-Induced Immune Disregulation in Mice

Mice that express the $H\text{-}2^s$; (H-2 is the mouse MHC) haplotype are also susceptible to $HgCl_2$ or gold salt-induced autoimmunity[37,38] however, clinical manifestations are milder than in rats. A.SW and SJL mice display an increase in serum IgE and IgG1 concentration and both strains develop a so-called immune complex type glomerulonephritis.[37, 38] Autoantibodies induced by $HgCl_2$ are mainly directed against nucleolar structures such as fibrillarin.[39] $HgCl_2$-injected BALB/c (H-2^d) mice also develop a glomerulonephritis and anti-nucleolar antibodies can be eluted from the kidneys of diseased mice.[40]

AUTOREACTIVE T CELL LINES

The role of T cells has been emphasized in the development of drug-induced autoimmunity in rats and in mice[19,41] and of immunosuppression in LEW rats,[32] however, the mechanisms at play might be different in rats and mice. In mice, it has been suggested that $HgCl_2$ or gold salts trigger a specific immune response against the drug or a self structure modified by the chemical.[42,43] In this case, chemicals would behave as haptens. Autoreactive T cells would be generated as a consequence of this specific immune response. It is also possible that these toxic agents modify the repertoire of self peptides constitutively expressed and allow an autoimmune response against normally cryptic determinants.[44]

We will focus on the mechanisms implicated in drug-induced disregulation in rats. $HgCl_2$ induces a polyclonal activation of BN T cells with a 2-fold increase in $CD4^+$ T cells at the peak of the disease (J. Fillon et al, in preparation); in LEW rats, $HgCl_2$ provokes a marked increase in the number of $CD8^+$ T cells.[32] These effects could be related to the ability of $HgCl_2$ to cross-link the TCR and to provoke an extensive protein tyrosine phosphorylation and an activation of the tyrosine kinase $p56^{lck}$.[45] This kinase is associated with CD4/CD8 and plays a role in T cell receptor-mediated signalling.[46] Among activated T cells, autoreactive T cells that recognize self MHC class II molecules or an unknown peptide constitutively expressed by MHC molecules have been detected in both strains with a frequency which peaks at the acme of the disease and then declines during the regulatory phase.[47,48] Autoreactive T cell lines have been derived from both strains treated with $HgCl_2$ or gold salts as follows: lymph node cells from $HgCl_2$-injected LEW rats or gold salt-injected BN rats were stimulated in the presence of syngeneic irradiated antigen presenting cells and expanded in IL-2 containing medium. After several months of stimulations performed in similar conditions every two weeks, T cell lines were stable; they were $CD4^+$ $CD8^-$ $TCR\alpha\beta^+$ and restricted by self Ia MHC class II molecules. T cell lines obtained from gold salt-injected BN rats (so-called BNAU lines) transferred the disease into normal BN recipients deprived of $CD8^+$ cells; this disease was even more severe than the active one. One of the T cell lines which was derived from LEW rats injected with $HgCl_2$ (so-called LEWHgA) was adoptively transferred into LEW rats which were subsequently challenged with myelin basic protein emulsified in complete Freund's adjuvant; the rats were completely protected against autoimmune encephalomyelitis.[47] The transfer of the LEWHgA T cell line 15 days before the first injection of $HgCl_2$ (given sc 100 μg/100 g bw three times a week) was also able to protect (LEWxBN) F1 hybrids against $HgCl_2$-induced autoimmunity (Bridoux et al, in preparation). In the first model, it has been shown that the autoreactive T cell line was able to recruit regulatory $CD8^+$ cells since depletion of $CD8^+$ cells abolished protection.[47] In the second model, depletion of $CD8^+$ cells did not prevent protection, which means that the regulation might be mediated by the line itself.

BNAU T cell lines and the LEWHgA T cell line differ by the profile of cytokines they produced upon stimulation. BNAU lines produce IL-4, no IL-2, no IL-10, no or little IFN-γ and are Th2-like cells.[49] LEW HgA produces some IL-2, IFN-γ, TGF-β, no IL-4 and an unidentified cytokine able to increase MHC class II molecules on B cells[47] (Bridoux et al, in preparation); this cytokine might be IL-10 but this remains to be determined.

Thus anti-class II T cell lines play a major role in both the development of autoimmunity and immunosuppression and these opposite effects are probably due to the differences in cytokines produced.

ROLE OF CYTOKINES IN $HGCL_2$ -INDUCED IMMUNE DISREGULATION

Disregulation of the Th1/Th2 Balance in $HgCl_2$-Induced Autoimmunity

A defect in the ability of BN rats to mount a Th1 response is probably partially at play in the susceptibility of this strain to $HgCl_2$-induced autoimmunity. P. Mathieson et al have shown that depletion of $OX22^{high}$ cells, a subset enriched in Th1 cells, aggravates the disease.[50] BN rats display a genetic defect in the production of IFN-γ and, in addition, $HgCl_2$ induces an early defect in the capacity of BN T cells to produce IFN-γ upon stimulation with ConA.[51] For example, BN rats exposed to inhaled antigens develop a strong IL-4 production in the absence of any production of IFN-γ while other strains also display an early production of IL-4, which is subsequently counterbalanced by IFN-γ.[52] The defect observed in BN rats could be due to a relative defect of CD8+ cells which are normally the main source of IFN-γ. It is also tempting to compare BN rats and BALB/c mice, also a Th2 prone strain that develops a glomerulopathy when injected with $HgCl_2$.[40] T cells from BALB/c mice transgenic for a TCR specific for a nominal antigen fail to respond to IL-12 when stimulated by the antigen in neutral conditions while B10 T cells which express the same transgene respond (Fig. 3.1).[53] By contrast, T cells from both strains respond to IL-12 when primary cultures are done in the presence of IL-12. This phenotype is under the control of one single gene.[53] It will be interesting to determine whether BN T cells present such a defect and to test whether the administration of IL-12 prevents $HgCl_2$-induced autoimmunity in BN rats.

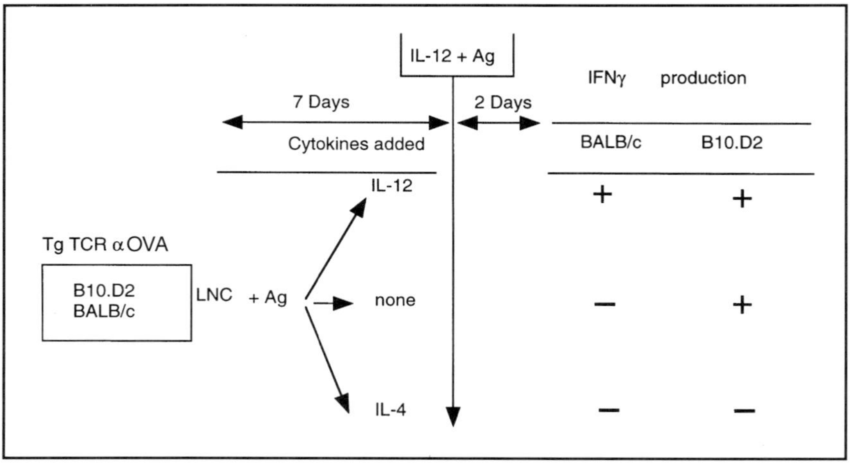

Fig. 3.1. Inability of T cells from BALB/c mice to respond to IL-12 in a secondary culture when the primary culture was done in the absence of exogenous cytokines. Naive T cells were obtained from BALB/c or B10.D2 mice that express a transgenic (Tg) T cell receptor (TCR) for ovalbumin (OVA). T cells were stimulated for 7 days by OVA either in the absence or in the presence of either IL-12 or IL-4. T cells were then restimulated by OVA in the presence of IL-12 and IFN-γ production was measured two days later.

We have treated BN with the OX81 anti-rat IL-4 mAb but it has not been possible to prevent autoimmune disease, probably because we were unable to neutralize $HgCl_2$-induced IL-4 even with high doses of the antibody (35 mg per rat) (S. Ramanathan, unpublished). The treatment of mice susceptible to $HgCl_2$-induced autoimmunity by an anti-IL-4 mAb abolishes the increase in serum IgE concentration and the production of IgG1 autoantibodies, but the titer of anti-nucleolar antibodies remains unchanged when compared to $HgCl_2$-injected mice.[54] Biancone et al have recently shown that inhibition of the interactions CD40-CD40L or CTLA4/CD28-B7s abrogates autoimmunity in $HgCl_2$-injected SJL mice without affecting IL-4 production.[55] The authors have also used anti-CD2 mAbs, which are able to immunosuppress Th1-mediated events, and have shown that these antibodies exacerbate autoantibody production, the production of IgE and the severity of the glomerulopathy.[55] The fact that gold salt-induced Th2 autoreactive T cell lines have to be in contact with normal BN B cells in order to detect autoantibody production in vitro shows that direct T-B interactions are also required for autoantibody production in rats.[49] All together these data strongly suggest that Th2 cells are responsible for the disease in rats and mice and that two events are required: (1) an activation of B cells mediated by direct interactions between B and T cells; and (2) IL-4 production.

Role of a Change in the Balance Th1/Th2 During the Regulatory Phase

Several phenomena probably participate in the improvement of the disease : the emergence of suppressor CD8+ cells and autoanti-idiotypic antibodies, and a decrease in the frequency of autoreactive T cells.[29,48,56] A change in the Th1/Th2 balance is probably also implicated and is supported by the fact that: (1)treating BN rats with an anti-IL-2 receptor mAb that affects Th1 cells but spares Th2 cells delays the regulation;[57] (2) mRNA for IL-12 may be detected in BN rats that improve (K. Gillespie, personal communication); (3) Th1-dependent functions such as IL-2 production and the generation of alloreactive cytotoxic T cells are improved when compared to those in normal BN rats.[58]

Cytokines at Play in $HgCl_2$-Mediated Immunosuppression

(LEWx BN) F1 hybrids injected with the LEWHgA T cell line and $HgCl_2$ 15 days later (100 μg/100g bw three times a week) do not develop $HgCl_2$-induced autoimmunity. The administration of a mouse anti-TGF-β mAb on day -3, 0, 6 and 9, with respect to the first injection of $HgCl_2$, completely abolishes the protection, which shows that this cytokine plays a prominent role in this regulatory process (Bridoux et al, in preparation). A role for this cytokine in the regulation of autoimmunity has been emphasized in several experimental-induced autoimmune diseases.[59] Moreover, it is also important in the physiological control of autoimmunity since knockout mice for this cytokine not only display multi-inflammatory organ disease,[60] but also a systemic lupus erythematosus and a Sjögren's syndrome.[61,62]

Effector mechanisms responsible for the resistance of LEW rats to $HgCl_2$-induced autoimmunity or for the LEWHgA-mediated protection towards EAE are still unknown. $CD8^+$ cell depletion induces very mild autoimmune manifestations with some renal IgM antibody deposition in $HgCl_2$-injected LEW rats and abrogates the protection against EAE in the LEW rats transferred with the LEWHgA T cell line. We have now evidence that the autoreactive LEWHgA T cell line is able to recruit $CD8^+$ and perhaps $CD4^+$ T cells and to convert them into regulatory T cells which may also produce TGF-β. From this point of view, it has been shown that the addition of TGF-β in the culture medium of a nephritogenic autoreactive $CD8^+$ T cell clone that normally produces high amounts of IFN-γ and no TGF-β converts it into a TGF-β producer line, unable to produce IFN-γ and devoid of nephritogenic potential.[63]

It is also possible that other cytokines such as IL-10, which would also be produced by LEWHgA T cells, are involved in the resistance of LEW rats to $HgCl_2$-induced autoimmunity. Such a hypothesis is supported by results of K. Gillespie et al who show that the IL-10 is the only cytokine upregulated in $HgCl_2$-injected LEW rats.[36] It is noteworthy that TGF-β may upregulate IL-10 production.[64] LEWHgA T cell type cells might represent a normal component of the immune response which is amplified by $HgCl_2$. These T cells could resemble autoreactive regulatory T cells described by Modigliani et al.[65,66] These latter T cells would be selected on thymic epithelial cells during neonatal life at a moment where negative selection on bone-marrow derived APCs would be not fully operative. These cells would recognize ubiquitous self peptides (as our LEWHgA T cell line), would be long-lived and would be able to convert recent thymic emigrants in regulatory cells specific for organ-derived peptides (Fig. 3.2). The mediator responsible for the recruitment of other regulatory cells is not IL-4; however, TGF-β has not yet been tested.[65,66]

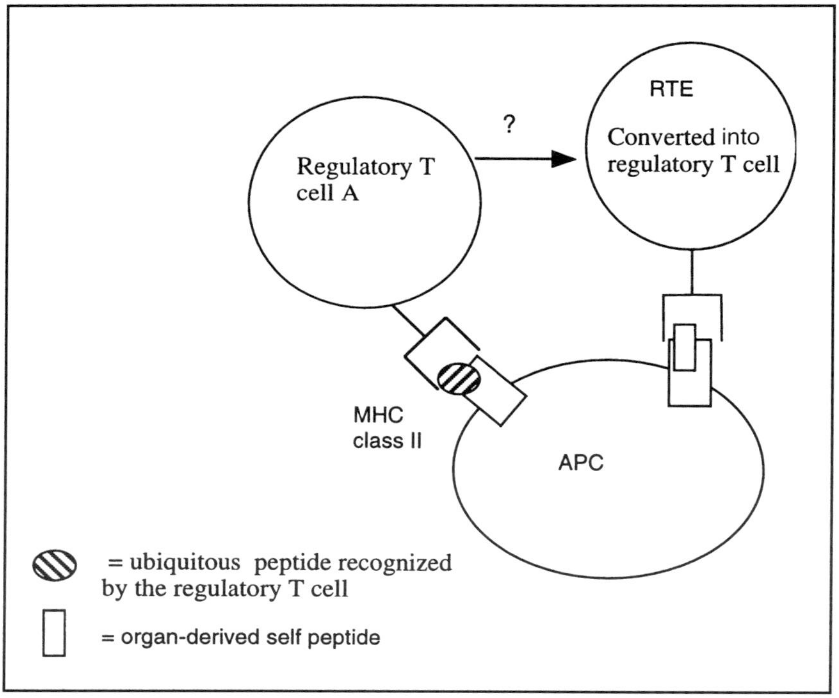

Fig. 3.2. Ability of regulatory T cells (called A on this figure) selected on thymic epithelium cells to convert organ-antigen specific recent thymic emigrant (RTE) cells into regulatory T cells. According to the hypothesis of Modigliani et al, the first regulatory T cell A recognizes an ubiquitous peptide (hatched circle) on an APC and educates the second one when the RTE cell recognizes its antigen (white square) on the same APC. If the RTE cell is stimulated by its antigen in the absence of A cells the RTE cell cannot be converted into a regulatory cell—it may just be inhibited by A cells.

HGCL$_2$-INDUCED IL-4 GENE TRANSCRIPTION

It has been recently shown that incubation of 98% pure T cells from normal BN rats with nontoxic amounts of $HgCl_2$ (5 μM) triggers IL-4 mRNA expression in less than 2 hours; $HgCl_2$ does not have this effect on LEW T cells.[67] Another group has also shown that $HgCl_2$ was able to induce IL-4 mRNA expression in BN but not in LEW mast cells.[68] The direct effect of $HgCl_2$ on the transcription of other cytokines such as IL-5, IL-10, IL-13 or TGF-β in BN and LEW T cells has not been reported. Addition of $HgCl_2$ to normal spleen cells from BN rats leads to an increase in MHC class II expression on B cells and an increase in CD23 expression, which supports the fact that $HgCl_2$ induces the production of IL-4.[67] This effect was indeed abrogated by coincubation with the anti-IL-4 mAb. Addition of $HgCl_2$ to normal LEW spleen cells is also associated with an increase in MHC class II expression on B cells, but without modification of the expression of CD23, which would suggest that $HgCl_2$ induces IL-10 production in LEW cells.[67]

Theoretically, as shown in Figure 3.3, $HgCl_2$ might act at the cell membrane level or downstream, for example by activating transcription factors and binding DNA which would favor the accessibility to RNA polymerase.[69] The effect of $HgCl_2$ was tested on murine T cell hybridomas in which ionomycin or ConA were able to induce IL-4 mRNA expression. $HgCl_2$ triggered mRNA IL-4 expression in some of these murine T cell hybridomas (A. Badou et al; in preparation). Therefore, we have used these cells as a tool for studying the mechanisms by which $HgCl_2$ acts. An inhibitor of transcription (actinomycin D) abolishes $HgCl_2$-mediated IL-4 gene transcription while an inhibitor of protein synthesis (cycloheximide) has no effect; this fits with the fact that $HgCl_2$ induces an early IL-4 gene transcription (A. Badou; in preparation). A chelator of intracellular calcium (BAPTA/AM) or an inhibitor of protein kinase C (H-7) inhibits IL-4 mRNA expression, which shows that the effect of $HgCl_2$ is Ca^{2+}- and PKC-dependent. This demonstrates that $HgCl_2$ does not act directly on transcription factors or at the promoter level but perhaps upstream at the cell membrane level. The regulation of IL-4 gene transcription mainly implies the fixation of transcription factors such as NF-AT and AP-1 on multiple elements of the promoter of the IL-4 gene; NF-AT is activated by an increase in intracellular calcium concentration and AP-1 by a pathway involving PKC,[70-75] (Fig. 3.3). Thus, a next step will be to look for an activation of these two factors upon $HgCl_2$ exposure.

The reasons why $HgCl_2$ triggers BN and not LEW T or mast cells to produce IL-4 are unknown; two hypotheses may be put forward: 1) the promoter of IL-4 gene might be polymorphic, the sequences of the BN haplotype allowing a better transcription of the gene. Such a polymorphism has been described in the human IL-4 gene promoter and some alleles are associated with a better

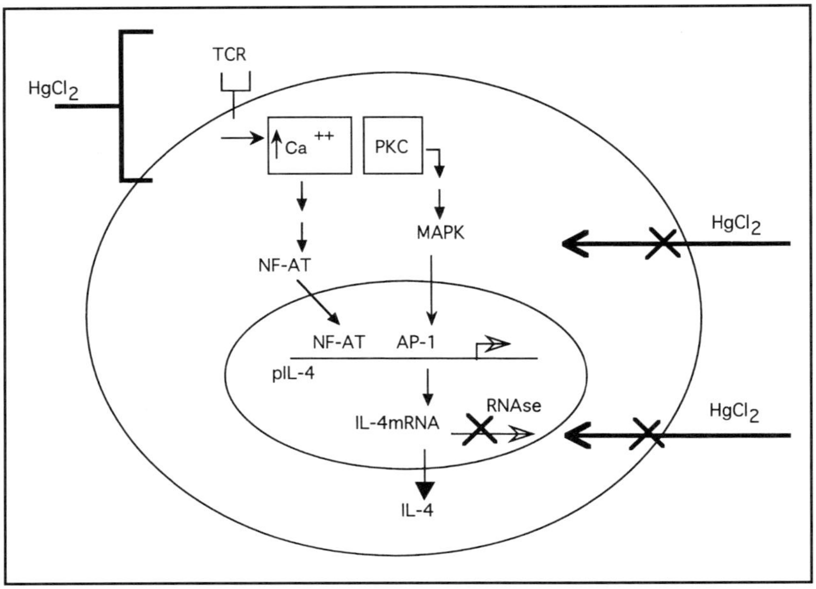

Fig. 3.3. Pathways leading to IL-4 gene transcription in a Th2 cell stimulated via its T cell receptor. Theoretical levels at which $HgCl_2$ might act. TCR = T cell receptor, MAPK = mitogen activated protein kinase, PKC = protein kinase C, NF-AT (nuclear factor of activated T cells) and AP-1 (activator protein-1) are transcription factors that play an essential role in IL-4 gene transcription. pIL-4 = promoter of IL-4 gene.

transcription of the gene.[76] A polymorphism between BN and LEW has already been described in the second intron of the gene.[77] The sequencing of BN and LEW IL-4 gene promoters is planned. The fact that one gene involved in the genetic susceptibility to gold salt-induced autoimmunity has been located on chromosome 10 in a region encompassing the gene encoding for IL-4 also supports this hypothesis.[77] However, it has to be mentioned that the IL-4 gene is functional in LEW rats, as has been shown in another experimental model.[14] A second hypothesis might also explain the reasons of the discrepancy: Th2 committed cells are three times more frequent in BN rats than in LEW rats based upon the differential expression of CD45RC molecules recognized by the OX22 mAb on T cells in both strains.[78,79] The ratio $OX22^{high}$ (enriched in Th1)/$OX22^{low}$ (enriched in Th2 cells) is 3:1 in LEW and 1:3 in BN rats. Normal LEW mast cells express less IgE receptor than BN mast cells[68] which might explain why LEW cells are insensitive to $HgCl_2$ if $HgCl_2$ interacts with these receptors. We now plan to purify $OX22^{low}$ and $OX22^{high}$ cells in the two strains and to assess the effects of $HgCl_2$ on each subset. We think that only $OX22^{low}$ cells from either strain would express IL-4 mRNA upon stimulation with $HgCl_2$ because they are already committed to become Th2 cells. Another way to determine whether $HgCl_2$ acts only on T cells engaged in Th2 differentiation will be to test the effects of $HgCl_2$ on Th1, Th2 and Th0 murine clones.

IMMUNE DEVIATION INDUCED BY HGCL$_2$ IN EAU

Experimental autoimmune uveoretinitis (EAU) is induced in susceptible LEW and (LEW x BN) F1 hybrids by immunization with retinal S antigen in adjuvants.[80] S-Antigen specific T cell lines and clones that produce IFN-γ and no IL-4 upon stimulation in vitro may adoptively transfer the disease into normal recipients. F1 rats that were susceptible to both Th1-mediated EAU and Th2-mediated $HgCl_2$-induced-autoimmunity were used in the experiments. They were injected with $HgCl_2$, and 7 days later (at a time where IL-4 production is induced) they were immunized with S-antigen in adjuvant. Rats were clinically and histologically protected against EAU. T cells from protected rats proliferated in the presence of S-antigen as well as those from control rats immunized with S antigen only; however, they produced different cytokines.[80] No or low amounts of IFN-γ were found in protected rats compared with diseased rats. By contrast, a strong expression of IL-4 mRNA was detected in protected rats and a weak one in diseased rats. T cells from protected or diseased rats were stimulated in vitro by S-antigen and were adoptively transferred into normal syngeneic hybrids. T cells from diseased rats transferred EAU while T cells from protected rats were quite unable to do so. The authors conclude that $HgCl_2$ induces the production of high amounts of IL-4, which conditions the differentiation of S-antigen-specific precursors in nonuveitogenic Th2 cells. It has been recently shown that low doses of IL-4 exacerbate IFN-γ production in vivo and in vitro and aggravate EAU;[81] treating (LEW x BN) hybrids with low doses of $HgCl_2$ has the same effect as low doses of IL-4.[81]

CONCLUSION

Anti-cytokine antibodies or cytokines may be used to treat autoimmune diseases. For example, antibodies directed against TNF-α, a proinflammatory cytokine, improve rheumatoid arthritis patients.[82] TGF-β, a cytokine with immunosuppressive properties, has also been used successfully used to treat experimental autoimmune diseases.[59] It is also conceivable to deviate a pathogenic Th1 response towards a nonagressive Th2 response as has been done successfully in some experimental autoimmune diseases (reviewed in ref. 6). Our models have allowed us to show that it is possible to modulate the immune response by using chemicals which are able to trigger cytokine production independently of a specific immune response. However, some difficulties may be encountered in trying to deviate the immune response:[80] First, IL-4, the cytokine responsible for Th2 differentiation, may promote Th1 cells when given at low doses.[81] Second, Th2 cells may be pathogenic and not regulatory as shown in our models.[49] They even may promote inflammatory reaction by recruiting other cells such as $\gamma\delta$ T cells.[83] By contrast, cells that produce IFN-γ may also produce immunosuppressive cytokines, IL-10 or TGF-β[84] (Bridoux et al, in preparation). Although controlling the Th1/Th2 balance offers interesting prospects in the treatment of autoimmune disorders, it would probably be more effective to induce regulatory cell subsets for therapeutical use.

References

1. Powrie F, Coffman RL. Cytokine regulation of T-cell function: potential for therapeutic intervention. Immunol Today 1993; 14:270-274.
2. Seder RA, Paul WE. Acquisition of lymphokine-producing phenotype by CD4+ T cells. Annu Rev Immunol 1994; 12:635-673.
3. Fowell D, McKnight AJ, Powrie F et al. Subsets of CD4+ T cells and their role in the induction and prevention of autoimmunity. Immunol Rev 1991; 123:37-64.
4. Romagnani S. Lymphokine production by human T cells in disease states. Annu Rev Immunol 1994; 12:227-257.
5. O'Garra A, Murphy K. Role of cytokines in determining T-lymphocyte function. Curr Opin Immunol 1994; 6:458-466.
6. Liblau RS, Singer SM, McDevitt HO. Th1 and Th2 CD4+ T cells in the pathogenesis of organ-specific autoimmune diseases. Immunol Today 1995; 16:34-38.
7. Ando DG, Clayton J, Kono D et al. Encephalithogenic T cells in B10.PL model of experimental allergic encephlomyelitis (EAE) are of the Th-1 lymphokine subtype. Cell Immunol 1989; 124:132-143.
8. Savion S, Grover S, Kawano Y et al. Uveitogenicity of T cell lines in the rat does not correlate with their production of TNF, INFγ or IL-3. ARVO. Investigative Ophthalmology & Visual Science 1992; 33:932.
9. Karpus WJ, Gould KE, Swanborg RH. CD4+ suppressor cells of autoimmune encephalomyelitis respond to T cell receptor-associated determinants on effector cells by interleukin-4 secretion. Eur J Immunol 1992; 22:1757-1763.
10. Karpus WJ. CD4+ suppressor cells inhibit the function of effector cells of experimental autoimmune encephalomyelitis through a mechanism involving transforming growth factor-β. J Immunol 1991; 146:1163-1168.
11. Chen Y, Inobe J-i, Weiner HL. Induction of oral tolerance to myelin basic protein in CD8-depleted mice: Both CD4+ and CD8+ cells mediate active suppression. J Immunol 1995; 155:910-916.
12. Chen Y, Kuchroo VK, Inobe J-i et al. Regulatory T cell clones induced by oral tolerance: suppression of autoimmune encephalomyelitis. Science 1994; 265:1237-1240.

13. Racke MK, Bonomo A, Scott DE et al. Cytokine-induced immune deviation as a therapy for inflammatory autoimmune disease. J Exp Med 1994; 180:1961-1966.
14. Saoudi A, Simmonds S, Huitinga I et al. Prevention of experimental allergic encephalomyelitis in rats by targeting autoantigen to B cells: evidence that the protective mechanism depends on changes in the cytokine response and migratory properties of the autoantigen-specific T cells. J Exp Med 1995; 182:335-344.
15. Rott O, Cash E, Fleischer B. Phosphodiesterase inhibitor pentoxifylline, a selective suppressor of T helper type1- but not type 2-associated lymphokine production, prevents induction of experimental autoimmune encephalomyelitis in Lewis rats. Eur J Immunol 1993; 23:1745-1751.
16. Racke MK, Burnett D, Pak S-H et al. Retinoid treatment of experimental allergic encephalomyelitis IL-4 production correlates with improved disease course. J Immunol 1995; 154:450-458.
17. Fossati L, Takahashi S, Merino R et al. An MRL/MpJ-lpr/lpr substrain with a limited expansion of lpr double-negative T cells and a reduced autoimmune syndrome. Int Immunol 1993; 5:525-532.
18. Singh RR, Kumar V, Ebling FM et al. T cell determinants from autoantibodies to DNA can upregulate autoimmunity in murine systemic lupus erythematosus. J Exp Med 1995; 181:2017-2027.
19. Goldman M, Druet P, Gleichmann E. T_H2 cells in systemic autoimmunity: insights from allogeneic diseases and chemically-induced autoimmunity. Immunol Today 1991; 12: 223-227.
20. Pelletier L, Pasquier R, Guettier C et al. $HgCl_2$ induces T and B cells to proliferate and differentiate in BN rats. Clin Exp Immunol 1988; 71:336-342.
21. Prouvost-Danon A, Abadie A, Sapin C et al. Induction of IgE synthesis and potentiation of anti-ovalbumin antibody IgE response by $HgCl_2$ in the rat. J Immunol 1981; 126:699-702.
22. Bellon B, Capron M, Druet E et al. Mercuric chloride-induced autoimmunity in Brown Norway rats : sequential search for anti-basement membrane antibodies and circulating immune complexes. Eur J Clin Invest 1982; 12:127-133.
23. Aten J, Bosman CB, Rozing J et al. Mercuric chloride-induced autoimmunity in the Brown Norway rat. Cellular kinetics and major histocompatibility complex antigen expression. Am J Pathol 1988; 133:127-138.
24. Druet E, Sapin C, Günther E et al. Mercuric chloride-induced anti-glomerular basement membrane antibodies in the rat. Genetic control. Eur J Immunol 1977; 7:348-351.
25. Pusey CD, Bowman C, Morgan A et al. Kinetics and pathogenicity of autoantibodies induced by mercuric chloride in the Brown Norway rat. Clin Exp Immunol 1990; 81:76-82.
26. Tournade H, Pelletier L, Guéry JC et al. Experimental gold-induced autoimmunity. Nephrol Dial Transplant 1991; 6:621-630.

27. Tournade H, Pelletier L, Pasquier R et al. D-penicillamine-induced autoimmunity in Brown Norway rats: similarities with $HgCl_2$-induced autoimmunity. J Immunol 1990; 144:2985-2991.
28. Mathieson PW, Thiru S, Oliveira DBG. Mercuric chloride-treated Brown Norway rats develop widespread tissue injury including necrotizing vasculitis. Lab Invest 1992; 67:121-129.
29. Mathieson PW, Stapleton KJ, Oliveira DBG et al. Immunoregulation of mercuric chloride-induced autoimmunity in Brown Norway rats: a role for CD8+ T cells revealed by in vivo depletion studies. Eur J Immunol 1991; 21:2105-2109.
30. Bigazzi PE. Lessons from animal model: the scope of mercury-induced autoimmunity. Clin Immunol Immunopathol 1992; 65:81-84.
31. Aten J, Veninga A, Bruijn JA et al. Antigenic specificities of glomerular-bound autoantibodies in membranous glomerulopathy induced by mercuric chloride. Clin Immunol Immunopathol 1992; 63:89-102.
32. Pelletier L, Pasquier R, Rossert J et al. $HgCl_2$ induces non specific immunosuppression in LEW rats. Eur J Immnol 1987; 17:49-54.
33. Pelletier L, Rossert J, Pasquier R et al. Effect of $HgCl_2$ on experimental allergic encephalomyelitis in Lewis rats. $HgCl_2$-induced downmodulation of the disease. Eur J Immunol 1988; 18:243-247.
34. Pelletier L, Galceran M, Pasquier R et al. Down-modulation of Heymann's nephritis by mercuric chloride. Kidney Int 1987; 32:227-232.
35. Gillespie KM, Qasim FJ, Tibbats LM et al. Interleukin-4 gene expression in mercury-induced autoimmunity. Scand J Immunol 1995; 41:268-272.
36. Gillespie KM, Saoudi A, Kuhn J et al. Th1/Th2 cytokine gene expression after mercuric chloride in susceptible and resistant rat strains. Eur J Immunol (in press).
37. Mirtcheva J, Pfeiffer C, Bruijn JA et al. Immunological alterations induced by mercury compounds. III. H-2A acts as an immune response and H-2E as an immune "suppression" locus for $HgCl_2$-induced antinucleolar autoantibodies. Eur J Immunol 1989; 19:2257-2261.
38. Pietsch P, Vohr HW, Degitz K et al. Immunopathological signs inducible by mercury compounds. II. $HgCl_2$ and gold sodium thiomalate enhance serum IgE and IgG concentrations in susceptible mouse strains. Int Arch Allergy Appl Immun 1989; 90:47-53.
39. Reuter R, Tessars G, Vohr H-W et al. Mercuric chloride induces autoantibodies against U3 small nuclear ribonucleoprotein in susceptible mice. Proc Natl Acad Sci USA 1989; 86:237-241.
40. Hultman P, Eneström S. Mercury-induced antinuclear antibodies in mice: characterization and correlation with renal immune complex deposits. Clin Exp Immunol 1988; 71:269-274.
41. Griem P, Gleichmann E. Metal ion-induced autoimmunity. Curr Opin Immunol 1995; 7:831-838
42. Schuhmann D, Kubicka-Muranyl M, Mirtschewa J et al Adverse immune reactions to gold: Chronic treatment with an Au(I) drug sensitizes T

cells not to Au(I), but to Au(III) and induces autoantibody formation. J Immunol 1990; 145:2132-2139.

43. Kubicka-Muranyi M, Griem P, Lubben B et al Mercuric chloride-induced autoimmunity in mice involves upregulated presentation by spleen cells of altered and unaltered nucleolar self antigen. Int Arch Allergy Immunol 1995; 108:1-10.
44. Griem P, Panthel K, Kalbacher H et al. Alteration of a model antigen by Au(III) leads to T cell sensitization to cryptic peptides. Eur J Immunol 1996; 26:279-287.
45. Nakashima I, Pu M-y, Nishizaki A et al. Redox mechanism as alternative to ligand binding for receptor activation delivering disregulated cellular signals. J Immunol 1994; 152:1064-1070.
46. Lee S-K, Shaw A, Maher SE, Bothwell ALM. $p59^{fyn}$ tyrosine kinase regulates $p56^{lck}$ tyrosine kinase activity and early TCR-mediated signaling. Int Immunol 1994; 6:1621-1627.
47. Castedo M, Pelletier L, Rossert J et al. Mercury-induced autoreactive anti-class II T cell line protects from experimental autoimmune encephalomyelitis by the bias of CD8+ anti-ergotypic cells in Lewis rats. J Exp Med 1993; 177:881-889.
48. Rossert J, Pelletier L, Pasquier R et al. Autoreactive T cells in mercury-induced autoimmunity. Demonstration by limiting dilution analysis. Eur J Immunol 1988; 18:1761-1766.
49. Saoudi A, Castedo M, Nochy D. et al. Self reactive anti-class II Th2 cell lines derived from gold salt-injected rats trigger B cell polyclonal activation and transfer autoimmunity in CD8-depleted normal syngeneic recipients. Eur J Immunol 1995; 25:1972-1979.
50. Mathieson PW, Thiru S, Oliveira DBG. Regulatory role of $OX22^{high}$ T cells in mercury-induced autoimmunity in the Brown Norway rat. J Exp Med 1993; 177:1309-1316.
51. van der Meide PH, de Labie MCDC, Botman CAD et al. Mercuric chloride down-regulates T cell interferon-γ production in Brown Norway but not in Lewis rats; role of glutathione. Eur J Immunol 1993; 23:675-681.
52. McMenamin C, Holt PG. The natural immune response to inhaled soluble protein antigens involves major histocompatibility complex (MHC) class I-restricted CD8+ T cell-mediated but MHC class II-restricted CD4+T cell-dependent immune deviation resulting in selective suppression of immunoglobulin E production. J Exp Med 1993; 178:889-899.
53. Jacobson NG, Szabo SJ, Güler ML et al. Regulation of interleukin-12 signal transduction during T helper phenotype development. Res Immunol 1995; 146:446-456.
54. Ochel M, Vohr H-W, Pfeiffer C et al. IL-4 is required for the IgE and IgG1 increase and IgG1 autoantibody formation in mice treated with mercuric chloride. J Immunol 1991; 146:3006-3011.

55. Biancone L, Andres G, Ahn H et al. Distinct regulatory roles of lymphocyte costimulatory pathways on T helper type 2-mediated autoimmune disease. J Exp Med 1996; 183:1473-1482.
56. Bowman C, Mason DW, Pusey CD et al. Autoregulation of antibody synthesis in mercuric chloride nephritis in the Brown Norway rat. I. A role for suppressor cells. Eur J Immunol 1984; 14:464-470.
57. Dubey C, Kuhn J, Vial MC et al. Anti-interleukin-2 receptor monoclonal antibody therapy supports a role for Th1-like cells in $HgCl_2$-induced autoimmunity in rats. Scand J Immunol 1993; 37:406-412.
58. Castedo M, Pelletier L, Pasquier R et al. Improvement of TH1 functions during the regulation phase of mercury disease in Brown Norway rats. Scand J Immunol 1994; 39:144-150.
59. Kuruvilla AP, Shah R, Hochwald GM et al. Protective effects of transforming growth factor β1 on experimental autoimmune diseases in mice. Proc Natl Acad Sci USA 1991; 88:2918-2921.
60. Shull MM, Ormsby I, Kier AB et al. Targeted disruption of the mouse transforming growth factor-β1 gene results in multifocal inflammatory disease. Nature 1992; 359:693-699.
61. Christ M, McCartney-Francis NL, Kulkarni AB et al. Immune dysregulation in TGF-β1-deficient mice. J Immunol 1994; 153:1936-1946.
62. Dang H, Geiser AG, Letterio JJ et al. SLE-like autoantibodies and Sjögren's syndrome-like lymphoproliferation in TGF-β knockout mice. J Immunol 1995; 155:3205-3212.
63. Meyers CM, Kelly CJ. Immunoregulation and TGF-β1. Suppression of a nephritogenic murine T cell clone. Kidney Int 1994; 46:1295-1301.
64. Maeda H, Shiraishi A. TGF-β contributes to the shift toward Th2-type responses through direct and IL-10-mediated pathways in tumor-bearing mice. J Immunol 1996; 156:73-78.
65. Modigliani Y, Pereira P, Thomas-Vaslin V et al. Regulatory T cells in thymic epithelium-induced tolerance. I. Suppression of mature peripheral non-tolerant T cells. Eur J Immunol 1995; 25:2563-2571.
66. Modigliani Y, Bandeira A, Coutinho A. A model for developmentally acquired thymus-dependent tolerance to central and peripheral antigens. Immunol Rev 1996; 149:156-174.
67. Prigent P, Saoudi A, Pannetier C et al. Mercuric chloride, a chemical responsible for Th2-mediated autoimmunity in Brown-Norway rats, directly triggers T cells to produce IL-4. J Clin Invest 1995; 96:1484-1489.
68. Oliveira DBG, Gillespie K, Wolfreys K et al. Compounds that induce autoimmunity in the Brown Norway rat sensitize mast cells for mediator release and interleukin-4 expression. Eur J Immunol 1995; 25:2259-2264.
69. O'Halloran TV. Transition metals in control of gene expression. Science 1993; 261:715-725.
70. Fraser JD, Straus D, Weiss A. Signal transduction events leading to T-cell lymphokine gene expression. Immunol Today 1993; 14:357-362.

71. Szabo SJ, Gold JS, Murphy TL et al. Identification of cis-acting regulatory elements controlling interleukin-4 gene expression in T cells: roles for NF-Y and NF-AT$_c$. Molec Cell Biol 1993; 13:4793-4805.
72. Todd MD, Grusby MJ, Lederer JA et al. Transcription of the interleukin 4 gene is regulated by multiple promoter elements. J Exp Med 1993; 177:1663-1674.
73. Rooney JW, Hodge MR, McCaffrey PG et al. A common factor regulates both Th1- and Th2-specific cytokine gene expression. EMBO Journal 1994; 13:625-633.
74. Rooney JW, Hoey T, Glimcher LH. Coordinate and cooperative roles for NF-AT and AP-1 in the regulation of the murine IL-4 gene. Immunity 1995; 2:473-483.
75. Rao A. NF-ATp: a transcription factor required for the co-ordinate induction of several cytokine genes. Immunol Today 1994; 15:274-281.
76. Song Z, Casolaro V, Chen R et al. Polymorphic nucleotides within the human IL-4 promoter that mediate overexpression of the gene. J Immunol 1996; 156:424-429.
77. Kermarrec N, Dubay C, De Gouyon B et al. Serum IgE concentration and other immune manifestations of treatment with gold salts are linked to MHC and IL-4 regions in the rat. Genomics 1996; 31:111-114.
78. Powrie F, Fowell D, McKnight AJ et al. 35th forum in immunology. CD4+ T-cell subsets: differentiation and function. Res Immunol 1991; 142:72-74.
79. Powrie F, Fowell D, McKnight AJ et al. Lineage relationships and functions of the CD4$^+$ T-cell subsets in the rat. Res Immunol 1991; 142:59.
80. Saoudi A, Kuhn J, Huygen K et al. TH2 activated cells prevent experimental autoimmune uveoretinitis, a TH1-dependent autoimmune disease. Eur J Immunol 1993; 23:3096-3103.
81. Ramanathan S, de Kozak Y, Saoudi A et al. Recombinant IL-4 aggravates experimental autoimmune uveoretinitis. J Immunol (in press).
82. Elliott MJ, Maini RN, Feldmann M et al. Treatment of rheumatoid arthritis with chimeric monoclonal antibodies to TNFα, safety, clinical efficacy and regulation of acute phase response. Arthritis Rheum 1993; 36:1681-1686.
83. Dieli F, Asherson GL, Colonna Romano G et al. IL-4 is essential for the systemic transfer of delayed hypersensitivity by T cell lines role of γδ cells. J Immunol 1994; 152:2698-2704.
84. Del Prete GF, De Carli M, Almerigogna F et al. Human IL-10 is produced by both type helper (Th1) and type 2 helper (Th2) T cell clones and inhibit their antigen-specific proliferation and cytokine production. J Immunol 1993;150:1-8.

CHAPTER 4

Immunoregulation of Experimental Autoimmune Diseases in the Rat

Don Mason, Francisco Ramírez, Victoria Heath,
Abdelhadi Saoudi, Ben Seddon and Phil Stumbles

INTRODUCTION

The Need for Mechanisms that Downregulate the Immune Response

A wide range of pathogens, from viruses to multicellular parasites, have the capacity to invade and replicate in mammals. Concurrently with the evolutionary adaptation of these pathogens to their hosts, the mammalian immune system has evolved a number of effector mechanisms to eliminate or at least control them. Classically these mechanisms are divided into innate and adaptive immunity. The essential difference between these is that the innate system is activated by pathogens such as bacteria that as a group are biochemically sufficiently unlike their mammalian hosts for them to be recognized as such, whereas the adaptive immune system requires specific recognition of minor biochemical differences between self and nonself. The adaptive immune system has arisen in evolution later than the innate one and, as would be expected, many of the effector molecules, such as interleukin-1 (IL-1), tumor necrosis factor (TNF) and complement are used by both. In addition, the adaptive immune system uses its own effector mechanisms such as antibody and cytotoxic T cells that are peculiar to it.

All of these mechanisms, both innate and adaptive, can be damaging to the mammalian host as well as to the pathogens that evoke them. Moreover, as described below, it is also possible for the host to fail to distinguish self from nonself, with the result that the immune system becomes autoreactive. These considerations imply the need for certain restraints upon the immune system if the host is not to suffer unduly from its response to foreign pathogens and to be nonresponsive to self antigens.

Immunointervention in Autoimmunity by Th1/Th2 Regulation,
edited by Luciano Adorini.

The Nature of the Restraining Regulatory Mechanisms

The restraints alluded to in the previous section influence both the innate and the adaptive immune systems and effect the induction phase, the nature of the immune response once elicited, its duration and its severity. Failures in any one of these aspects of immunological control can result in disease. The relevant check points are briefly as follows:

(i) An immune response should be elicited only by stimuli that are potentially pathogenic. Thus foreign antigens introduced in such a way as to present no threat to the host do not activate the armamentarium of immunological effector mechanisms, but instead are capable of inducing tolerance.[2,3] It is evident that the phylogenetically older innate immune system plays a critical role in determining whether or not the adaptive immune system is called into action and this fact explains how the mode of presentation of a foreign antigen can determine whether it elicits an immune response or induces specific nonreactivity.[4] It also explains how an adjuvant, which is quite unrelated to a particular antigen, can potentiate an immune response to it and can also influence the nature of that response. For example, complete Freund's adjuvant (CFA), which contains heat-killed mycobacteria, induces via the innate immune system, the release of IL-12 from macrophages and this cytokine drives the differentiation of T cells towards those that mediate delayed-type hypersensitivity reactions. Conversely, the adjuvant alum promotes antibody responses and the synthesis of IgE.

(ii) An immune response, once evoked, must be limited in its spatial and temporal extent and this condition requires that mechanisms exist that antagonize or neutralize this response. Several of these regulators are soluble forms of cytokine receptors. The soluble receptor for IL-1 (sIL-1R)[5] is such an example and several others have been described. Cytokines themselves play a major role in controlling immune responses : IL-4, IL-10 and IL-13, for example, all downregulate cell-mediated immune responses[6-8] while IFN-γ and IL-12 promote them.[9,10] Similarly, many of the molecules in the complement system exist to limit its activation or the effects of its activation.[11]

An interesting incidental consequence of the development by the immune system of mechanisms that modulate the immune response is that pathogenic organisms are known to exploit the existence of such immunological antagonists. Vaccinia has captured the mammalian gene for sIL-1R[12] and the Epstein-Barr virus has incorporated into its genome the gene for the anti-inflammatory cytokine IL-10.[13] There are many other examples.

(iii) A limit must be imposed upon the vigor of an immune response. It is not commonly recognized that the neuroendocrine system plays a vital role in this regard. Rats and mice that are adrenalectomized die within 48 hours of being immunized with complete Freund's adjuvant.[14,15] In the intact animal the injection of CFA induces the production of IL-1 and TNF as part of the innate response. The amount of the highly toxic TNF that is generated is subject to a negative feedback mechanism: the IL-1 released into the blood stream activates the hypothalamic pituitary adrenal axis by inducing the discharge of corticotrophin releasing factor into the hypothalamic-pituitary venous portal system.[16,17] As a consequence the corticosterone secreted by the adrenal glands as a consequence provides sufficient suppression of TNF production to prevent it from causing unregulated toxicity. Adrenalectomized animals cannot use this negative feedback mechanism and die of toxic shock.[18] The acute vulnerability of patients with autoimmune destruction of the adrenal glands (Addison's disease) to otherwise nonlethal infections can probably be attributed to the same mechanism.

Later in this chapter evidence will be provided that in addition to the acute protective effects of the adrenal glands in modulating immune responses the endogenous synthesis of corticosteroids can have a long-lasting effect on the adaptive immune response.

(iv) The adaptive immune response must be capable of distinguishing self from nonself. Autoimmune diseases represent a failure of such discrimination and it is evident that any discussion of the pathogenesis of these diseases must be based on an understanding of the physiological mechanisms that allow immune responses to foreign antigens but not to self.

The remainder of this chapter reviews briefly what is known of these mechanisms and gives an account of our own experimental findings.

The Problem of Self/Nonself Discrimination by the Adaptive Immune System

As already indicated, foreign antigens that do not themselves activate the innate immune system and that are presented in the absence of adjuvants may induce tolerance rather than immunity. However, there is no reason to suppose that self antigens, in the presence of an ongoing immune response to a pathogen, are intrinsically tolerogenic. The B cell and T cell antigen receptor repertoires are generated by rearrangements of DNA segments drawn from a large pool of genetic elements. The selection of these genetic elements is random and it follows that the receptor repertoires so produced do not, at the outset, distinguish self from nonself : such distinction is an acquired characteristic of the immune system. It has been shown that during their differentiation both B cells and T cells pass through a transient phase in which contact with their specific antigen leads to clonal deletion and this mechanism for purging the lymphocyte repertoire of autoreactive cells is of major importance in establishing self/nonself discrimination.[19,20] However, it is clearly imperfect, as is evidenced by the existence of a diversity of autoimmune diseases. Genetically determined differences in the degree to which clonal deletion is achieved provide, in principle, an explanation for why such conditions develop in some individuals but not in others and this hypothesis may explain, for example, why the second susceptibility locus for insulin-dependent diabetes mellitus (IDDM) in the human (IDDM-2) maps to the promoter region of the insulin gene.[21,22] However, this explanation does not account for the fact that there are many other susceptibility loci for IDDM both within and outside the major histocompatibility locus (MHC) and it is evident that clonal deletion is but one mechanism for achieving self tolerance.[23]

In the following section data are described that indicate that self tolerance is in part maintained by an active immunological process that prevents autoimmunity.

Self Tolerance as a T Cell Mediated Feature of the Immune System

It has been demonstrated that normal healthy laboratory rats (and, by extension, humans too), that do not develop any autoimmune disease, harbor T cells that are potentially capable of causing insulin-dependent diabetes (IDDM) and autoimmune thyroiditis.[24,25] These animals remain free of disease because their autoreactive T cells are under the control of some regulatory mechanism that is itself T cell-mediated (ref. 25 and see below). It follows that the existence of self-reactive T cells does not necessarily lead to autoimmune disease and while the failure of clonal deletion to completely purge the T cell repertoire of autoreactive cells plays an important part in autoimmunity it is not principally at this level that the pathogenesis of autoimmunity must be sought.

There is evidence that T cells that prevent autoimmunity are specific for the self antigens for which they express their protective function (reviewed in ref. 26). If this conclusion is valid, and it is difficult to find an alternative explanation for the data, then an apparent paradox exists in that the T cells that prevent autoimmunity are themselves autoreactive. However, the study of T cell subpopulations, defined either in terms of the repertoire of cytokines that they secrete on activation or by cell surface phenotype, has established that both the $CD4^+$ and $CD8^+$ T cell types are functionally heterogeneous and this property provides a possible solution to the paradox described above.[27-32] In the mouse $CD4^+$ T cell clones can be allocated into one of three groups depending on whether they produce IL-4, IL-5, IL-10 and IL-13 (Th2 clones) or TNF, IFN-γ and IL-2 (Th1 clones) or a mixture of these cytokines (Th0 cells).[27] Human T cells can be similarly classified.[31] These functional specializations are not some artifact of T cell cloning since rat (reviewed in refs. 29 and 30) and mouse[32] peripheral $CD4^+$ T cells can be subdivided into those that predominantly produce Th1 or Th2 type cytokines. The subdivision is made on the basis of which CD45 isoforms are expressed: in the rat $CD4^+$ T cells that express the C isoform of CD45 at high levels produce more IL-2 and IFN-γ than those that are CD45RC^{low} . Conversely, these latter cells are high IL-4 producers on activation.[29,30] Mouse CD4$^+$ T cells separated according to their level of expression of CD45RB are broadly similar to their rat counterparts with the expression of CD45RB in the mouse correlating with CD45RC in the rat.[32]

The significance, for the control of autoimmunity, of this functional specialization among CD4 T cells is that Th1 cells and Th2 cells appear to be mutually antagonistic. For example, IL-4, produced by Th2 clones, inhibits the differentiation of Th1 cells and IFN-γ, produced by these latter cells, prevents the induction of Th2 responses.[27] These interactions help to explain why antibody synthesis and delayed type hypersensitivity reactions, requiring respectively the induction of Th2 and Th1 responses, tend to be reciprocally expressed in vivo.[33] Further, the in vitro and in vivo observations on T cell antagonism have lead to the hypothesis that cell-mediated autoimmune diseases, such as IDDM, experimental allergic encephalomyelitis (EAE) and Hashimoto's thyroiditis might be normally prevented by cells that make a Th2 type response to the relevant autoantigens.[34,35] Evidently autoimmune diseases that are antibody-mediated, such as autoimmune hemolytic anemia and myasthenia gravis, cannot be controlled in this way as Th2 type responses would be expected to promote, rather than inhibit, antibody synthesis. In addition there are, perhaps, few autoimmune diseases that are caused purely by antibody or cell-mediated reactions. Despite this apparent difficulty it is well established that for some autoimmune diseases at least, T cells play an essential role in their control and there is evidence (see below) that these cells can inhibit both types of autoimmune effector mechanism.

In the remainder of this chapter some of the principles outlined above are further developed using specific experimental examples to illustrate them.

STUDIES ON THE CONTROL OF AUTOIMMUNE DISEASES IN THE RAT

Autoimmune Diabetes Induced by Perturbation of the Immune System

The active participation of the immune system in preventing autoimmunity is well illustrated by the induction of IDDM in rats. Thymectomy of rats as weanlings or young adults, followed by 1000 Rad ^{137}Cs γ–irradiation administered in four equal doses at two week intervals results in the development of autoimmune diabetes in virtually all the males and some 70% of females.[25,30] Disease starts as early as three weeks after the last dose of irradiation and the majority of animals become diabetic within the ensuing seven weeks. If the rats in the prediabetic phase are treated with a monoclonal antibody that depletes CD8$^+$ T cells they are completely protected from diabetes. However, the injection of $CD8^+$ cells from nontreated healthy syngeneic donors results in the reappearance of the disease. It is evident that these donor animals possess T cells that have the potential to cause IDDM but that this potential is held in check by some radiation sensitive control mechanism.

The nature of this mechanism was revealed by the finding that a particular subset of $CD4^+$ T cells from normal donors could prevent the diabetes if injected into prediabetic recipients after the last dose of γ-irradiation. Protection was complete and required 5 x10^6 cells with the phenotype $CD4^+$ $CD45RC^{low}$ TCR $\alpha\beta^+$ $RT6^+$. These are cells with a memory phenotype and they are resistant to long-term thymectomy. The mechanism of protection and the antigen specificity of these cells is not known but is currently being studied. However, because the protective lymphocytes had the characteristics of primed T cells it was anticipated that $CD4^+$ $CD8^-$ thymocytes, being an unprimed population, would not be as potent as the peripheral cells in preventing IDDM. Contrary to this expectation this thymocyte subset was very effective. As few as 6 x 10^5 protected 50% of recipients from diabetes whereas peripheral $CD4^+$ $CD45RC^{low}$ TCR $\alpha\beta^+$ cells failed to prevent diabetes in any recipient if used at doses of less than almost 10 times this number. Despite this potency, increasing the thymocyte dose to 5 x 10^6 cells did not prevent disease in all recipients and this failure is likely to be a reflection of the fact that the $CD4^+$ $CD8^-$ thymocyte subset contains both cells that can prevent diabetes and those that can cause it.[36] This conclusion is supported by the observation that unfractionated $CD4^+$ peripheral T cells also fail to confer protection from diabetes on more than about 50% of animals. It is known that the $CD45RC^{high}$ subset of these cells contains ones that are autoreactive so that the outcome with regard to the development of diabetes of the transfer of peripheral $CD4^+$ cells involves a conflict between two subpopulations, one that is diabetogenic and one that is protective. This observation suggests that a mechanism acts in the periphery to ensure that only protective T cells enter the $CD4^+$ $CD45RC^{low}$ T cell pool. The nature of this mechanism is not known.

While it is evident that much remains to be determined about the active control of T cells with autoimmune potential several conclusions can be drawn from the above data. The selection of the T cell repertoire in the thymus generates a population of cells that is particularly potent at preventing diabetes (and presumably other organ-specific autoimmune diseases). However, this selection procedure also permits the emigration of potentially autoreactive T cells into the periphery. Finally, the peripheral $CD4^+$ T cells that prevent diabetes in the rat are phenotypically distinguishable from those that are involved in the pathogenesis of the disease.

These conclusions suggest that rats that develop IDDM following thymectomy and γ-irradiation do so because they lose the protection mediated by $CD4^+$ $CD45RC^{low}$ T cells. It may be that IDDM that arises spontaneously in the human, in the NOD mouse and in the BB rat results from a failure of this mechanism.

The Prevention of EAE by Autoreactive T Cell Lines Producing Th2 Type Cytokines

In the preceding section it was suggested that the development of autoimmune diabetes was prevented in normal animals by the action of self-reactive $CD4^+CD45RC^{low}$ T cells that produced cytokines that inhibited the activation of a damaging Th1 T cell response to β cells in the pancreas. The following account shows that a protective mechanism can be specifically induced by the appropriate activation of autoreactive T cells in another autoimmune disease, experimental allergic encephalomyelitis (EAE).

EAE is an animal model of autoimmune disease that can be induced in the appropriate animal strain after immunization with myelin basic protein (MBP) in CFA; this is termed active EAE. The disease can also be induced after the intravenous injection of MBP-specific $CD4^+$ T cells into naive recipients, and this is designated as passive EAE. EAE in Lewis rats is a monophasic disease characterized by a single episode of paralysis followed by spontaneous recovery and animals that have recovered are refractory to attempts to induce further episodes of disease. EAE has been studied as a model of multiple sclerosis (MS) in humans and, although the validity of EAE as a model of MS has been argued, it is a useful system in which to study the regulation of the immune and inflammatory responses at different levels.[38] Understanding the immune response during EAE may help to interpret some aspects of MS, for example the spontaneous recovery shown by some animals with EAE may be similar to the acute remissions seen in some MS patients.

The mechanisms responsible for the spontaneous recovery of EAE and the maintenance of the refractory phase, although objects of intense research, are not completely understood. Various mechanisms have been proposed including suppressor T lymphocytes,[39,40] suppressor macrophages,[41] anti-T lymphocyte idiotype responses,[42] serum suppressor factors,[43] production of immuosuppressive factors by glial cells,[44] regulation by IFN-γ,[45] regulation by IL-10,[46,47] by TGF-β[48] and neuroendocrine-mediated immunoregulation.[15,49] It has been shown that endogenously produced corticosterone plays an essential role in the spontaneous recovery of rats from EAE: adrenalectomized animals do not recover spontaneously but instead develop an unremitting and progressive paralysis with a fatal result. Adrenalectomized rats given corticosterone replacement therapy at the onset of disease recover within a few days and develop the refractory state like nonadrenalectomized rats.[15] These observations have lead to the conclusion that the spontaneous recovery of rats from EAE occurs because corticosteroids, which are found at high levels in the sera of paralyzed animals, acutely depress the autoimmune response through their immunosuppressive and anti-inflammatory effects. However, once animals have recovered from EAE, serum steroid levels return to normal and adrenalectomy after recovery does not induce relapse of disease, which indicates that corticosteroids are not directly responsible for the maintenance of the refractory phase of the disease.[15] This conclusion does not exclude the possibility that the transiently elevated levels of corticosterone during paralysis have long-term effects on the immune response to MBP and, relevant to this issue, it has been suggested that glucocorticoids induce not only an acute inhibitory action on the immune response but also mediate some long-lasting effects which affect the balance between cell-mediated immunity and humoral immunity.[50,51]

To explore this possibility a protocol was used that was similar to the one employed by others to study the effects of cytokines on the development of the immune response. The experimental scheme is outlined in Figure 4.1. Rat $CD4^+$ T cells were activated in vitro with the polyclonal mitogen Concanavalin A (Con A) in the presence or absence of the glucocorticoid analogue dexamethasone (DEX). They were then expanded in IL-2, again some cultures in the presence of DEX, and subsequently restimulated, this time in the absence of the hormone. Cytokine gene expression and total T cell recovery was evaluated following this secondary activation. The results, summarized semiquantitatively in Table 4.1, indicate that the exposure of the cells to DEX in the primary stimulation changed the cytokine synthesis induced by the secondary stimulation. A marked decrease in IFN-γ and TNF production was observed and this was associated with an equally marked increase in IL-4 protein release and in mRNAs for IL-10 and IL-13. Glucocorticoid exposure did not significantly affect IL-2 and IL-6 production nor the total numbers of T cells recovered. Similar results were obtained when the T cells were stimulated in a mixed lymphocyte reaction (MLR) rather than by the use of Con A. These data show that, in addition to their acute immunosuppressive effects, glucocorticoids may display a more subtle and long-lasting influence on the immune response, inducing in $CD4^+$ T lymphocytes a Th2 pattern of cytokine production on reactivation.[53]

The significance of these results with respect to the refractory phase of EAE in the rat is that the disease is a cell-mediated one involving T cells that produce IL-2 and IFN-γ[54] and IL-4 producing MBP-specific lines are not able to induce passive EAE (see below). Similar findings have been described in the mouse EAE model: MBP-specific Th1 clones can induce disease but Th2 clones do not and, in some circumstances, can inhibit the development of EAE.[55,56] Based on these observations the refractory phase of EAE can in principle be explained if endogenously produced glucocorticoids induce a switch from a Th1 to a Th2 response to MBP.[50] The sequence of events may be as follows: MBP

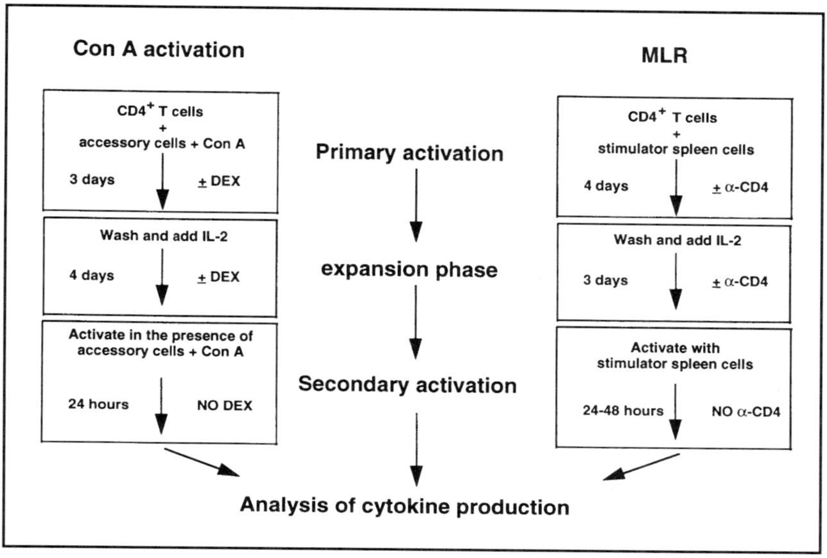

Fig. 4.1. Experimental scheme to study the effect of glucocorticoids on cytokine production by $CD4^+$ cells.

immunization with CFA induces T cells that synthesize inflammatory cytokines and these migrate to the CNS and cause nerve conduction defects with resultant paralysis. However, the concomitant stress activates the HPA axis and corticosterone is released which arrests the disease through its acute anti-inflammatory effects. In addition the long-term effects of glucocorticoids on T cell differentiation will give rise to a Th2 type response that will prevent the further activation of disease-inducing MBP-reactive Th1 cells.

This hypothesis provides an explanation for the recovery and refractory periods, although a detailed analysis of the cytokines produced during the different phases of EAE, and how endogenous glucocorticoids affect this synthesis, is necessary to verify it. However, the hypothesis does suggest that it might be possible to manipulate the immune response to MBP in such a way as to induce resistance to the induction of EAE and experiments that illustrate that such manipulation is achievable will now be described.

Effects of IL-4 and Glucocorticoids on MBP-Specific T Cell Lines

In this section are described some new results obtained analyzing the effects of glucocorticoids and IL-4 on cytokine production and in the generation of MBP-specific T cell lines. These cell lines do not transfer passive EAE but are very effective at inducing resistance to the active induction of EAE with MBP in CFA.

Table 4.1. Effects of dexamethasone on cytokine gene expression and cellular proliferation[a]

Cytokine	Culture conditions	
	Con A	**Con A + DEX**
IFN-γ	++	+
TNF	++	+
IL-4	-[b]	++
IL-10	+	++
IL-13	+	++
IL-2	+	+
IL-6	+	+
Proliferation [c]	+	++

[a] The production of cytokines is expressed in arbitrary units according to the differences obtained with protein assay except for IL-10 and IL-13 that were analysed by RT-PCR due to the lack of a protein assay.
[b] IL-4 mRNA was detected by RT-PCR but no protein was detected by the class II induction assay.
[c] Proliferation is expressed in arbitrary units.

In an extension of the work described in the previous section a comparison was made of the effects of IL-4 and DEX on cytokine production.[53] A synergy between IL-4 and DEX was observed in the generation of a Th2 response in that cells treated with both reagents together produced high levels of IL-4 and very low levels of IL-2, IFN-γ and TNF, which is the characteristic Th2 phenotype.[27] This protocol has now been applied to the generation of Th2 type MBP-reactive T cell lines in order to study their effects on the induction of EAE. Lewis rats were immunized with MBP/CFA and draining lymph nodes used as a source of MBP-reactive T cells for in vitro derivation of cell lines. In the first round of activation in vitro with MBP the cultures were supplemented with DEX, IL-4 or both reagents together. The subsequent rounds of activation were performed in the absence of these reagents. The protocol to obtain the antigen-specific lines is depicted in Figure 4.2. It was found that IL-4 and DEX together promoted the development of MBP Th2 lines more effectively than either IL-4 or DEX alone and cell lines were obtained that produced high amounts of IL-4 but low levels of TNF and IFN-γ. When IL-4 and DEX were omitted from the primary in vitro stimulation typical Th1 type lines were obtained. The results are briefly summarized in Table 4.2.

Consistent with other published work[55,56] it was observed that these Th-2 lines could not transfer passive EAE and, in contrast to the results obtained

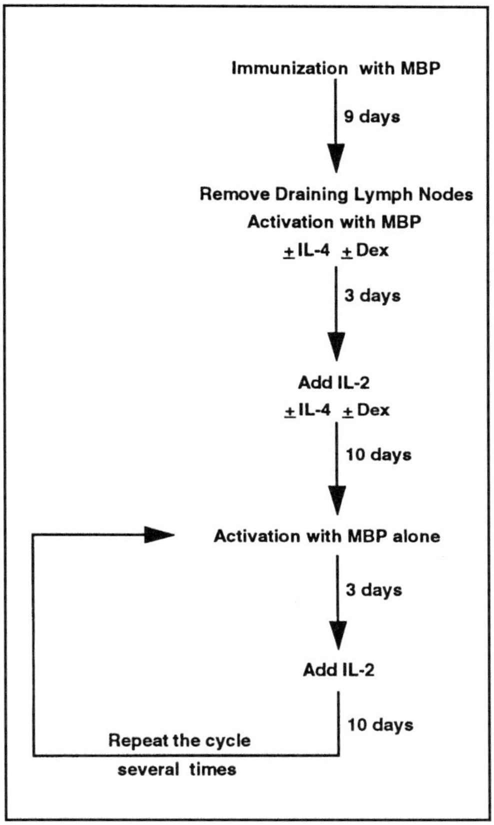

Fig. 4.2. Protocol to obtain MBP-specific Th1 and Th2 cell lines.

with encephalitogenic Th1 lines, very few of the Th2 cells infiltrated the nervous system (Table 4.2). These data suggest an association between the production of cytokines and the ability to home to tissues. A similar observation was obtained in rats protected from EAE by injection of the encephalitogenic peptide conjugated to anti-rat IgD. Cells from these protected animals elaborate a Th2 response to MBP and do not accumulate in the CNS.[57] In the current experiments it was found that the MBP-specific Th2 lines were unable to inhibit the induction of passive EAE by Th1 MBP-specific lines: when the two types of T cell line were injected together the animals developed EAE. This result indicated that Th2 cells could not control disease induction by preactivated Th1 cells but gave no indication of their possible ability to prevent de novo induction of such encephalitogenic cells. To answer this question experiments were carried out to determine whether animals injected with MBP-specific Th2 lines developed EAE when immunized with MBP/CFA. Animals were injected with

Table 4.2. Effect of IL-4 and DEX on the ability to transfer passive EAE and regulate the response to MBP by MBP-specific T cell lines[a]

MBP lines	Cytokine production IFN-γ	IL-4	Passive EAE	CNS homing	Active EAE	Humoral response
Control lines	High	Very low	+	+	+/-	+
IL-4 + DEX treated lines	Low	High	-	-	-	-

[a]Animals were injected with MBP-specific T cell lines and examined for signs of disease (passive EAE). Some of the animals were sacrificed for histological analysis of the spinal cord. Two or four weeks later the remaining animals were immunized with MBP in CFA (active EAE). One month after the MBP immunization animals were bled and the sera were used to analyse anti-MBP antibodies.

encephalitogenic and nonencephalitogenic lines and after two to four weeks the animals were immunized with MBP in CFA. It was found that animals injected with Th1 type cell lines developed passive EAE as expected but were also relatively resistant to the subsequent induction of active EAE. In contrast, recipients of the Th2 type cell lines did not develop passive EAE and were completely resistant to active EAE (Table 4.2). At present we do not have a full explanation for the ability of Th1 lines, transferred into naive recipients to give partial protection against the subsequent induction of active EAE, but it is known that the transfer of such cells also induces corticosterone release which plays an essential part in the control of passive EAE.[15] Analysis of the antibody responses to MBP in the animals given the Th1 or Th2 type T cell lines gave the unexpected result that there was inhibition of anti-MBP humoral responses in animals injected with the IL-4 and DEX treated lines, but not in animals injected with Th1 lines (Table 4.2). We do not yet know the mechanism that promotes the inhibition of the antibody response to MBP by the lines derived from IL-4 and DEX treated cells, but the result calls into question whether these cells are truly Th2 lines.

The Use of Anti-CD4 Monoclonal Antibodies In Vivo to Prevent EAE

In both of the preceding sections on the control of autoimmunity it is apparent that $CD4^+$ T cells play an essential role in preventing the activation of other T cells that are potentially autoreactive. However, in experiments where autoimmunity is prevented or arrested by the use of anti-CD4 monoclonal antibodies used as therapeutic agents it might be supposed that these antibodies mediate their beneficial effects simply by depleting recipients of T cells with this phenotype. Given the central role that CD4+ T cells play in cell-mediated immunity this interpretation of the demonstrable efficacy of these reagents is attractive. As the following data demonstrate, this view is certainly not always correct: some anti-CD4 monoclonal antibodies that prevent autoimmunity do so without depleting recipients of CD4+ cells. This result, while encouraging in as much as these reagents are less likely to be as destructive to the immune system as are depleting antibodies, raises the question of how they mediate their beneficial effects. The following describes what is known of the answer.

The demonstration that anti-CD4 monoclonal antibodies could virtually completely inhibit the MLR in vitro[58] prompted a study of the in vivo effects of these antibodies. It was found that the anti-rat CD4 antibody W3/25, which blocked the rat MLR, could completely prevent the induction of EAE in Lewis rats immunized with guinea-pig MBP in CFA. In these initial experiments the antibody was given before the onset of the disease, but it was further shown that if the antibody was given at the onset of paralysis the progression of the disease could be rapidly arrested.[59] Within 24 hours of such treatment recovery from disease was virtually complete.

While these acute effects could be reasonably attributed to the ability of the antibody to prevent T cell activation, as evidenced by its effect on the MLR, an unexpected result was that rats treated with the antibody to prevent EAE did not develop the disease after the injected antibody had been metabolized by the recipients. Measurements of the numbers of CD4+ T cells in these animals indicated that these were substantially normal so that the continued non-responsiveness in vivo could not be attributed to any gross T cell depletion. Further, CD4+ T cells from these disease-free rats could be stimulated in vitro with MBP and the resultant T cell blasts, when injected intravenously into normal Lewis rat recipients, induced severe EAE.[60] These findings demonstrated that MBP-reactive T cells had been primed in the donor animals and had not been depleted by the anti-CD4 antibody.

These observations lead to a study of the long-term effects of activating rat CD4+ T cells in vitro in the presence of the W3/25 antibody.[61] The experimental protocol was similar to that already described in the studies on the effects of glucocorticoids on cytokine gene expression described in the preceding section (see Fig. 4.1). However, the nonspecific T cell mitogen Con A was not used but instead T cells were activated by allogeneic irradiated stimulator cells. Mixed leukocyte cultures (MLC) were set up, using purified CD4+ responder T cells and heavily irradiated allogeneic splenocytes as stimulators, in the presence or absence of the antibody. After four days some of the culture supernatants were assayed for IL-2, IL-4 and IFN-γ. The responder cells from these primary MLCs were washed and expanded in IL-2 for a further three days in the absence of the antibody. After this period the cells were again washed and restimulated for 48 hours in a secondary MLC using irradiated stimulators as before. Measurements of cytokine synthesis were made following this secondary activation by RT-PCR (for IL-4 and IL-13), by bioassays (for IL-2 and IL-4) and by ELISA (for IFN-γ synthesis).

The data showed that the introduction of the monoclonal antibody into the primary MLC had, as expected, a potent inhibitory effect on T cell proliferation although at high responder cell numbers, $>10^6$ cells/well, this reached almost 50% of that seen in the control cultures. The presence of the antibody had marked effects on cytokine synthesis in these primary cultures: IL-2 synthesis was reduced by 70% but IFN-γ production was completely eliminated (Table 4.3). (No IL-4 was detected in any of the primary cultures). The reduced proliferation in the antibody-containing cultures was not influenced by the addition of recombinant IL-2 but an unexpected result was the finding that adding recombinant IL-4 caused vigorous T cell proliferation, indicating that the cells were fully responsive to this cytokine. Proliferation of the control cultures, that is those not containing the anti-CD4 antibody, was not changed by the addition of recombinant IL-2 or IL-4.

The effects of including the anti-CD4 antibody in the primary MLC on cytokine synthesis in the secondary one were striking. Whereas the control and experimental cultures produced IL-2 and IFN-γ in similar amounts only the experimental cultures generated large amounts of the Th2 type cytokines IL-4 and IL-13 (Table 4.3).[61]

These results obtained in vitro suggest an explanation for the long-term in vivo effects of the W3/25 monoclonal antibody on EAE, specifically that in the presence of the antibody, MBP-reactive T cells are induced to develop a Th2 type cytokine repertoire. As reported in the previous section it has been shown that CD4⁺ T cell lines that transfer EAE into naive recipients produce IL-2, TNF and IFN-γ indicating that this disease is caused by Th1 type T cells. Consequently the deviation of the response to a Th2 type one by the antibody would be expected to prevent disease. However, it must be noted that such a deviation cannot be complete because, as reported above, $CD4^+$ T cells from antibody protected rats immunized with MBP/CFA can, after in vitro stimulation with MBP, adoptively transfer EAE into recipient animals. This finding suggests that in rats in which EAE has been prevented by the injection of the nondepleting anti-CD4 monoclonal antibody W3/25, encephalitogenic T cells are primed but are actively prevented from inducing disease by other MBP-reactive T cells that produce cytokines that inhibit Th1 type responses. There are clear similarities between this conclusion and those discussed earlier in this article on the control of IDDM by $CD4^+$ T cells injected into lymphopaenic rats and on the

Table 4.3. In vitro effects of W3/25 (anti-CD4) monoclonal antibody on cytokine synthesis in the primary and secondary MLC

	W3/25 in primary culture	**IL-2**	**IFN-γ**	**IL-4**
Primary MLC	-	30	100	NF*
	+	10	0	NF
Secondary MLC	-	40	500	1.1
	+	50	550	NF

Measurements of IL-2, IFN-γ and IL-4 are expressed in Units/ml based on titrations using recombinant proteins as standards. Assays of IFN-γ and IL-4 were made 24 hours after stimulation and IL-2 after 48 hours.*NF, None found.

prevention of the induction of EAE by the injection of MBP-reactive cells programmed in vitro by dexamethasone and IL-4.

CONCLUSION

In the three experimental systems described above it is evident that potentially autoreactive T cells can be prevented from causing disease by other T cells that control rather than delete them. As noted earlier in this article there is now a substantial amount of data that demonstrate that self tolerance cannot be accounted for solely in terms of clonal deletion and T cell anergy. The data from the experiments on lymphopaenia-induced diabetes illustrate this fact and demonstrate the ability of the immune system to suppress unwanted immune responses by generating regulatory T cells of a defined phenotype. While the antigen-specificity of these cells is unknown the fact that they arise spontaneously and that there appears to be an intrathymic mechanism for their generation strongly suggests that such regulatory cells provide a natural, physiological mechanism for the prevention of autoimmunity. If this is so then it may ultimately prove possible to prevent or suppress autoimmune diseases in the human by stimulating this protective population. One striking feature of the incidence of IDDM in the spontaneously diabetic NOD mouse is that it is higher in animals kept in a specific pathogen free environment and it may be that the rising level of IDDM in the human may also be associated with a reduction in the range of environments that organisms encounter in modern hygienic conditions. One feature of the immune system that was mentioned at the beginning of this article was the need for the immune system to satisfy two conflicting demands, to respond vigorously to pathogens but also to regulate the magnitude of the response. The immune system has evolved in conditions more akin to those in the Third World than to those in developed countries and the many polymorphisms among genes that encode the immune response have undergone selective pressures that are quite different to those generated by modern Western environments. It may be, as the data from the NOD mice suggest, that environmental antigens, particularly enteric ones, play a role in stimulating the regulatory T cells that prevent autoimmunity.

In this review we have described experimental manipulations that can control and even totally prevent experimentally induced IDDM and EAE. How this protective function is mediated is not presently understood but it seems evident that self tolerance is in part an active process involving regulatory T cells whose specificity, at least in EAE, is for the self antigen to which a potentially damaging autoreactive response is directed. The findings that the cells that prevent EAE on passive transfer to MBP-immunized recipients appear to have a Th2 like cytokine repertoire might be interpreted to indicate that the protective mechanism in these experiments depends on the well-known antagonism between Th2 and Th1 responses. However, there is one observation that is incompatible with this explanation. In the experiments on EAE, whether involving in vivo treatment of rats with anti-CD4 mAbs or the in vitro conditioning of MBP-reactive T cells with dexamethasone and IL-4, the prevention of EAE is associated with a reduced titer of antibody to MBP. This finding indicates that the disease-preventing protocols have inhibitory effects on both cell-mediated responses to MBP *and* humoral ones. There is evidence that TGF-β plays an essential role in the control of EAE in mice[48] and it may be that it is this cytokine that is involved in the experiments that we describe. Further work is required to establish whether the Th2 cytokines such as IL-4 and IL-13 also play some role in the T cell mediated control of autoimmunity but if they do their ability to promote antibody responses must be inhibited in some way. In any event it would seem unlikely that a physiological mechanism for T cell-mediated control of autoimmunity would function satisfactorily if it simply induced a Th2 response to a self antigen in place of a Th1 type one. Despite these questions that remain to be resolved, the current data provide strong support for the concept of active control of autoimmunity and there are clear therapeutic implications in this conclusion.

References

1. Matzinger P. Tolerance, danger, and the extended family. Annu Rev Immunol 1994; 12:991-1045.
2. Guéry JC, Galbiati F, Smiroldo S et al. Selective development of T helper (Th)2 cells induced by continuos administration of low dose soluble proteins to normal and β2-microglobulin-deficient BALB/c mice. J Exp Med 1996; 183:485-497.

3. Romball CG, Weigle WD. In vivo induction of tolerance in murine $CD4^+$ cell subsets. J Exp Med 1993; 178:1637-1644.
4. Fearon DT, Locksley RM. The instructive role of innate immunity in the acquired immune response. Science 1996; 272:50-54.
5. Symons JA, Young PR, Duff GW. Soluble type II interleukin-1 (IL-1) receptor binds and blocks processing of IL-1 beta precursor and loses affinity for IL-1 receptor antagonist. Proc Natl Acad Sci USA 1995; 92:1714-1718.
6. Powrie F, Coffman R. IL-4 and IL-10 inhibit DTH and IFN-γ production. Eur J Immunol 1993; 23:2223-2229.
7. Fiorentino DF, Zlotnik A, Vieira P et al. IL-10 acts on the antigen-presenting cell to inhibit cytokine production by Th1 cells. J Immunol 1991; 146:3444-3451.
8. Doherty TM, Kastelein R, Menon S et al. Modulation of murine macrophage function by IL-13. J Immunol 1993; 151:7151-7160.
9. Gajewski TF, Joyce J, Fitch FW. Antiproliferative effect of IFN-γ in immune regulation III Differential selection of TH1 and TH2 murine helper T lymphocyte clones using recombinant IL-2 and recombinant IFN-γ. J Immunol 1989; 143:15-22.
10. Trinchieri G. Interleukin-12: a proinflammatory cytokine with immunoregulatory functions that bridge innate resistance and antigen-specific adaptive immunity. Annu Rev Immunol 1995; 13:251-276.
11. Campbell RD, Law SKA, Reid KBM et al. Structure, organization, and regulation of the complement genes. Annu Rev Immunol 1988; 6:161-195.
12. Alcami A, Smith GL. A soluble receptor for interleukin-1β encoded by vaccinia virus: a novel mechanism of virus modulation of the host response to infection. Cell 1992; 71:153-167.
13. Moore K, Vieira P, Fiorentino D et al. Homology of cytokine synthesis inhibition factor (IL-10) to Epstein-Barr virus gene BCRF1. Science 1990; 248:1230-1234.
14. Bertini R, Bianchi M, Ghezzi P. Adrenalectomy sensitizes mice to the lethal effects of interleukin 1 and tumor necrosis factor. J Exp Med 1988; 167:1708-1712.
15. MacPhee IAM, Antoni FA, Mason DW. Spontaneous recovery of rats from experimental allergic encephalomyelitis is dependent on regulation of the immune system by endogenous adrenal corticosteroids. J Exp Med 1989; 169:431-445.
16. Besedovsky H, Sorkin E, Keller M et al. Changes in blood hormone levels during the immune response. Proc Soc Exp Biol Med 1975; 150:466-470.
17. Blalock JE. A molecular basis for bidirectional communication between the immune and neuroendocrine systems. Phisiol Rev 1989; 69:1-32.
18. Munck A, Guyre P, Holbrook NJ. Physiological functions of glucocorticoids in stress and their relation to pharmacological actions. Endocrin Rev 1984; 5:25-44.

19. Goodnow CC. Transgenic mice and analysis of B cell tolerance. Annu Rev Immunol 1992; 10:489-518.
20. Kappler JW, Roehm M, Marrack P. T cell tolerance by clonal elimination in the thymus. Cell 1987; 49:273-280.
21. Bennett ST, Lucassen AM, Gough SCL et al. Susceptibility to human type 1 diabetes at IDDM2 is determined by tandem repeat variation at the insulin gene minisatellite locus. Nature Genetics 1995; 9:284-292.
22. Kennedy GC, German MS, Rutter WJ. The minisatellite in the diabetes susceptibility locus IDDM2 regulates insulin transcription. Nature Genetics 1995; 9:293-298.
23. Wicker LS, Todd JA, Peterson LB. Genetic control of autoimmune diabetes in the NOD mouse. Annu Rev Immunol 1995; 13:179-200.
24. Penhale WJ, Farmer A, Irvine WJ. Thyroditis in T cell depleted rats. Influence of strain, radiation dose, adjuvants and antilymphocyte serum. Clin Exp Immunol 1975; 21:362-375.
25. Fowell D, Mason DW. Evidence that the T cell repertoire of normal rats contains cells with the potential to cause diabetes. Characterization of the CD4$^+$ T cell subset that inhibits this autoimmune potential. J Exp Med 1993; 177:627-636.
26. Saoudi A, Seddon B, Heath V et al. The physiological role of regulatory T cells in the prevention of autoimmunity: the function of the thymus in the generation of the regulatory T cell subset. Immunol Rev 1996; 149:195-216.
27. Mosmann TR, Coffman RL. TH1 and TH2 cells: different patterns of lymphokine secretion lead to different functional properties. Annu Rev Immunol 1989; 7:145-173.
28. Carter LL, Dutton RW. Type 1 and Type 2 T-cell subsets. Curr Op Immunol 1996; 8:336-342.
29. Powrie F, Mason D. Phenotypic and functional heterogeneity of CD4$^+$ T cells. Immunol Today 1988; 9:274-277.
30. Fowell D, McKnight AJ, Powrie F et al. Subsets of CD4$^+$ T cells and their roles in the induction and prevention of autoimmunity. Immunol Rev 1991; 123:37-64.
31. Romagnani S. Human Th1 and Th2: doubt no more. Immunol Today 1991; 12:256-257.
32. Bottomly K. Phenotype and functional heterogeneity of CD4$^+$ T cells. Immunol Today 1988; 9:268-274.
33. Parish CR, Liew FY. Immune response to chemically modified flagellin III Enhanced cell-mediated immunity during high and low zone antibody tolerance to flagellin. J Exp Med 1972; 135:298-311.
34. Charlton B, Lafferty KJ. The Th1/Th2 balance in autoimmunity. Curr Op Immunol 1995; 7:793-798.
35. Day MJ, Tse AGD, Puklavec M et al. Targeting autoantigen to B cells prevents the induction of a cell-mediated autoimmune disease in rats. J Exp Med 1992; 175:655-659.

36. Saoudi A, Seddon B, Fowell D et al. The thymus contains a high frequency of cells that prevent autoimmune diabetes on transfer into prediabetic recipients. Submitted 1996.
37. Powrie F, Mason D. OX-22^{high} $CD4^+$ T cells induce wasting disease with multiple organ pathology: prevention by the OX-22^{low} subset. J Exp Med 1990; 172:1701-1708.
38. Zamvil SS, Steinman L. The T lymphocyte in experimental allergic encephalomyelitis. Annu Rev Immunol 1990; 8:579-621.
39. Welch AM, Swanborg RH. Characterisation of suppressor cells involved in regulation of experimental allergic encephalomyelitis. Eur J Immunol 1976; 6:910-912.
40. Ellerman KE, Powers JM, Brostoff SW. A suppressor T-lymphocyte cell line for autoimmune encephalomyelitis. Nature 1988; 331:265-267.
41. Welch AM, Swierkosz JE, Swanborg RH. Regulation of self tolerance in experimental allergic encephalomyelitis. I. Differences between lymph node and spleen suppressor cells. J Immunol 1978; 121:1701-1705.
42. Lider O, Reshef T, Beraud E et al. Anti-idiotypic network induced by T cell vaccination against experimental allergic encephalomyelitis. Science 1988; 239:181-183.
43. Paterson PY, Harwin SM. Suppression of allergic encephalomyelitis in rats by mean of antibrain serum. J Exp Med 1963; 117:755-774.
44. Fontana A, Kristensen F, Dubs R et al. Production of prostaglandin E and an interleukin-1 factor by cultured astrocytes and C6 glioma cells. J Immunol 1982; 129:2413-2419.
45. Billiau A, Heremans H, Vanderkerckhove F et al. Enhancement of experimental allergic encephalomyelitis in mice by antibodies against IFN-γ. J Immunol 1988; 140:1506-1510.
46. Kennedy MK, Torrance DS, Picha KS et al. Analysis of cytokine mRNA expression in the central nervous system of mice with experimental autoimmune encephalomyelitis reveals that IL-10 mRNA expression correlates with recovery. J Immunol 1992; 149:2496-2505.
47. Issazadeh S, Ljungdahl A, Hojeberg B et al. Cytokine production in the central nervous system of Lewis rats with experimental autoimmune encephalomyelitis: dynamics of mRNA expression for interleukin-10, interleukin-12, cytolysin, tumor necrosis factor alpha and tumor necrosis factor beta. J Neuroimmunol 1995; 61:205-212.
48. Santambrogio L, Hochwald GM, Saxena B et al. Studies on the mechanisms by which transforming growth factor-beta (TGF-beta) protects against allergic encephalomyelitis. Antagonism between TGF-beta and tumor necrosis factor. J Immunol 1993; 151:1116-1127.
49. Levine S, Sowinski R, Steinetz B. Effects of experimental allergic encephalomyelitis on thymus and adrenal: relation to remission and relapse. Proc Soc Exp Biol Med 1980; 165:218-224.
50. Mason D. Genetic variation in the stress response: susceptibility to experimental allergic encephalomyelitis and implications for human inflammatory disease. Immunol Today 1991; 12:57-60.

51. Daynes RA, Araneo BA, Dowell TA et al. Regulation of murine lymphokine production in vivo. III. The lymphoid tissue microenviromeot exerts regulatory influences over T helper cell function. J Exp Med 1990; 171:979-996.
52. Swain SL, McKenzie DT, Weinberg AD et al. Characterization of T helper 1 and 2 cell subsets in normal mice. Helper T cells responsible for IL-4 and IL-5 production are present as precursors that require priming before they develop into lymphokine-secreting cells. J Immunol 1988; 141:3445-3455.
53. Ramirez F, Fowell D, Puklavec M et al. Glucocorticoids promote a Th2 cytokine response by $CD4^+$ T cells in vitro. J Immunol 1996; 156:2406-2412.
54. Sedgwick JD, MacPhee IAM, Puklavec M. Isolation of encephalitogenic $CD4^+$ T cell clones in the rat. Cloning methodology and interferon-γ secretion. J Immunol Methods 1989; 121:185-196.
55. Chen Y, Kuchroo VK, Inobe J et al. Regulatory T cell clones induced by oral tolerance: suppression of autoimmune encephalomyelitis. Science 1994; 265:1237-1240.
56. Kuchroo VK, Das MP, Brown JA et al. B7-1 and B7-2 costimulatory molecules activate differentially the Th1/Th2 developmental pathways: application to autoimmune disease therapy. Cell 1995; 80:707-718.
57. Saoudi A, Simmonds S, Huitinga I et al. Prevention of experimental allergic encephalomyelitis in rats by targeting autoantigen to B cells: evidence that the protective mechanism depends on changes in the cytokine response and migratory properties of the autoantigen-specific T cells. J Exp Med 1995; 182:335-344.
58. Webb M, Mason DW, Williams AF. Inhibition of mixed lymphocyte response by monoclonal antibody specific for a rat T lymphocyte subset. Nature 1979; 282:841-843.
59. Brostoff SW, Mason DW. Experimental allergic encephalomyelitis: successful treatment in vivo with a monoclonal antibody that recognizes T helper cells. J Immunol 1984; 133:1938-1942.
60. Sedgwick JD, Mason DW. The mechanism of inhibition of experimental allergic encaphalomyelitis in the rat by monoclonal antibody against CD4. J Neuroimmunol 1986; 13:217-232.
61. Stumbles P, Mason D. Activation of $CD4^+$ T cells in the presence of a non-depleting monoclonal antibody to CD4 induces a Th2-type response in vitro. J Exp Med 1995; 182:5-13.

CHAPTER 5

Immune Deviation Resulting from the Action of Regulatory T Cells or from the Nature of the Antigenic Determinant Can Control Autoimmunity

Vipin Kumar and Eli Sercarz

Self-tolerance to tissue specific antigens reflects at least in part an active dynamic state in which potentially pathogenic, self-reactive T cells are prevented from causing disease by other, regulatory T cells. The differential secretion of lymphokines by T cell subsets offers an explanation for the ability of certain T cells to induce autoimmunity and others to regulate these autoreactive T cells.[1-3] Antigen-specific CD4 T cell responses can be generally categorized as type 1 (Th1) or type 2 (Th2).[4] Th1 cells secrete IL-2, IFN-γ and TNF-β, whereas Th2 cells secrete IL-4, IL-5 and IL-10. Type 1 effector functions include macrophage activation, and involvement in B cell differentiation to produce IgG2a antibodies important for opsonization and phagocytosis. Th1 cells are involved in host resistance to pathogens, pathogen-induced inflammation and autoimmunity, and tissue damage. Type 2 effectors are mainly involved in B cell differentiation leading to the secretion of IgG1 or neutralizing antibodies and IgE, important for mast cell degranulation, and eosinophil activation. Notably, cross-regulation enables one or the other phenotype to dominate in several antigenic systems. Modulatory type 2 cytokines, for example IL-4 and IL-10, are crucial for the inhibition of inflammatory responses. Similarly, IFN-γ inhibits expansion of Th2 effectors.

Type 1 cytokines have been shown to be related to several cell-mediated autoimmune diseases. For example, IFN-γ, lymphotoxin or TNF-α secretion

---

Immunointervention in Autoimmunity by Th1/Th2 Regulation, edited by Luciano Adorini.

correlate with the encephalitogenic capacity of T cell clones reactive to myelin basic protein (MBP).[5] Anti-TNF-α antibodies have been shown to inhibit experimental autoimmune encephalomyelitis (EAE) and collagen-induced arthritis while anti-TGF-β accelerates these diseases (reviewed in refs. 1 and 2). It seems that both TGF-β and IL-10 play an important role in regulating autoimmune inflammatory reactions. However, it is important to emphasize that generalized effects of different cytokines are unpredictable, owing to the temporal and localized induction of cytokines during an immune response.

Conditions leading to the spontaneous in vivo activation of either self-reactive T cells with particular effector functions, for example MBP-reactive T cells secreting IFN-γ and lymphotoxin, or regulatory TCR-peptide-specific T cells in B10.PL mice, are poorly understood. MBP-reactive type 1 T cell lines and clones, capable of inducing EAE in naive animals upon adoptive transfer, can be established following in vivo and in vitro priming. There are several examples where in vivo priming under appropriate conditions can result in T cells predominantly secreting IL-4/IL-5 or type 2 cytokines, thereby resulting in protection from autoimmune diseases (reviewed in refs. 1 and 2). Regulatory T cells could be specific for the antigen itself (for example, antigen-specific type 2 T cells) or for the TCR displayed on disease-causing T cells.[3] Here, we will discuss different settings in which the structure of the antigenic determinant, or the nature of the activated TCR repertoire, influences the direction of T cell development and thus autoimmunity. We will also discuss our work which shows that immune deviation in regulatory populations can affect the engagement of the regulatory T cell network during the course of EAE in B10.PL or (SJL x B10.PL)F1 mice. It is crucial to permit the regulatory T cells to exercise their program, which appears to be dependent upon the specific cytokine milieu.

THE STRUCTURE OF ANTIGENIC T CELL DETERMINANTS AND IMMUNE DEVIATION

Flanking Regions of A Determinant Can Give Rise to a Type 1 or a Type 2 Predominant Response

It has been demonstrated that T cells reactive to multiple determinants on MBP or even other myelin proteins become activated during the course of the autoimmune response.[6-8] It has also been suggested that recruitment of T cells against subdominant determinants may be responsible for relapses in the chronic relapsing model of EAE in SJL/J mice.[8] There exists the possibility that recruitment of T cells specific for other determinants on the same protein or on different proteins may ameliorate disease if these newly recruited T cells are of the Th2 type and secrete regulatory cytokines like IL-4/IL-10 or TGF-β. A critical aspect of the mutually inhibitory cytokine system is that any initial skewing of the response will be perpetuated in favoring that population which was originally activated.

Experimental approaches presently tested in these systems have utilized synthetic peptides to activate T cells in vivo or in vitro. Since the choice of peptide is arbitrary, in some cases a Th1-inducing, and in others a Th2-inducing, direction will be initiated *simply by the choice of the peptide.* In the absence of concrete knowledge about which members of the envelope of naturally processed and presented peptides is employed predominantly, and considering the fact that residues upstream or downstream of the core determinant can be crucial in influencing Th1 or Th2 polarization, it becomes important to discover which of the determinants is actually responsible for the initiation of autoaggression. Thus, MBP-reactive T cells can be expanded towards the Th1 phenotype and induce EAE, but this does not necessarily prove that such T cells are the natural instigators of EAE. We have carefully studied the response to an MBP determinant in the region between 121-150 in B10.PL mice and noticed that depending upon the choice of truncated peptide, one can either generate a Th1 or a Th2 predominant response (Kumar et al, unpublished observations). Thus, B10.PL mice immunized with 121-140 contract EAE, whereas those immunized with 131-150 do not. The central determinant appears to lie between residues 131-140 and T cells specific for 121-140 or 131-150 are cross-reactive. Therefore, dependent upon the predominant, naturally processed determinants and which most closely resemble the peptide chosen by the investigator, the T cells thus induced may be either protective or encephalitogenic. In this case, it appears that the naturally primed T cells prefer the subdominant determinant, with downstream flanking residues (131-150) that render poor binding to class II A^u. Thus, MBP121-140 binds >500-fold better to $I\text{-}A^u$ than MBP131-150. Notably, 131-150-specific naturally primed T cells predominantly secrete IL-4, consistent with our recent findings related to MHC-binding affinity and lymphokine secretion in response to peptide variants of MBP Ac1-9.[8] Since these cells are Th2-like, it is likely that they are protective and modulate the course of the natural response to MBP. Indeed, B10.PL mice vaccinated with 131-150 are significantly protected from Ac1-9-induced EAE. In contrast, mice vaccinated with 121-140 contract severe and chronic EAE (Kumar, unpublished observations). This finding is consistent with the scenario that following monophasic disease induction with Ac1-9, the dominant determinant of MBP in B10.PL mice, determinant spreading to other subdominant regions of MBP may not enhance disease but actually limit it. Thus, the outcome of determinant spreading may ultimately be determined by the balance between proinflammatory Th1 and protective Th2 responses to multiple determinants on the same and/or different myelin proteins.

The Available TCR Repertoire Can Be Decisive in Controlling Type 1 or 2 Predominance

T cells derived from B10.PL or PL/J mice of the H-2^u haplotype predominantly utilize TCR Vβ8.2 or Vβ13 in combination with only Dβ2 and Jβ2 gene segments for the recognition of the dominant determinant, Ac1-9/A^u, of myelin basic protein.[10,11] NZW mice, with identical class II H- 2^u genes (I-A and I-E), carry an 8.8 kb deletion in their TCR β-chain locus encompassing Dβ2 and Jβ2 gene segments.[12] How does this deletion of the crucial Dβ2-Jβ2 region in NZW mice influence specific responses to Ac1-9/A^u as well as to other known A^u or E^u determinants of MBP? We found that these mice respond very poorly to the dominant Ac1-9/A^u and to the sub-dominant 31-50/E^u determinant in T cell proliferation assays or in the induction of experimental autoimmune encephalomyelitis (EAE).[13] This loss of response apparently can be attributed to the absence of high avidity TCRs with essential CDR3 residues contributed by Dβ2 or Jβ2 gene segments. These data demonstrate constraints in the recognition of certain antigenic structures in spite of the myriad possibilities for TCR diversity. Although the presence of extremely low affinity/frequency T cells of Ac1-9/A^u or 35-47/E^u-specificity in NZW mice cannot be ruled out, their activity is not detectable with in vitro (proliferative response) or in vivo (EAE-induction) assays. Recently, we found a low frequency of Ac1-9-specific T cells in NZW mice predominantly secreting IL-4. Thus, presence of low affinity type 2 T cells must be tested in each instance. These cells usually do not proliferate well and are likely to be poorly encephalitogenic. Thus, it is likely that holes in the repertoire do exist, probably due to the absence of TCRs employing appropriate Dβ-Jβ gene segments. It seems that the interaction between TCR structures and their ligands, Ac1-9/A^u or 31-50/E^u complexes, is stringent enough to demand specific residues in the CDR3 region (Dβ2-Jβ2), despite the availability of closely related gene segments within the Dβ1-Jβ1 region.[13]

IgH Congenics Are Differentially Susceptible to EAE and Show a Type 1 or 2 Bias

The induction of EAE is associated both with particular MHC and non-MHC alleles. In certain susceptible haplotypes, such as H-2^s, strains with particular non-MHC genes, manage to avoid EAE. SJL/J (Igh^b) mice are susceptible while A.SW (Igh^e)are not susceptible to MBP- induced EAE, although both of these strains express the susceptible MHC haplotype, H-2^s. We have studied the lymph node T cell proliferative responses to MBP in these and other H-2^s mice and found that disease-resistant strains appear to be poor responders. It is interesting that although A.SW mice respond poorly to whole MBP, the central immunodominant determinant of MBP in the H-2^s haplotype (MBP 81-100) induces very efficient proliferative responses as well as EAE.[14] Our data suggest that the poor proliferative responses to MBP may correlate with either the absence of Igh-b or the presence of the Igh-e locus in some of the congenic strains that we have tested (Kumar, unpublished observations). It is possible that the Igh locus may affect the developing T cell repertoire through idiotypic interactions and may lead to the skewing of a subpopulation of MBP-reactive cells towards the Th2 phenotype. This becomes more relevant in the context of the demonstration that different mouse strains can respond to the same determinant and give rise to clones that produce either IFN-γ or IL-4.[9,14,15] Therefore if EAE were only induced by Th1 clones, it is plausible that in a nonsusceptible strain, IL-4 producers might preferentially expand and indirectly protect the animals from EAE.[14,16] Preliminary data suggest that in mice with the Igh^e locus, the initial antigen-specific response is dominated by Type 2 cells but this is not the case in mice with the Igh^b locus (Kumar, unpublished).

Class II MHC-Binding Affinity of the Determinant Can Result in Type 1 or Type 2 Bias

Both Th1 and Th2 cells differentiate from a common pool of precursors postthymically and as a consequence of activation by the antigen.[17] Cytokines produced by Th1 and Th2 cells play an important role in mutual regulation of their differentiation and subsequent effector functions.[4] Do interactions among the MHC/peptide, TCR and other accessory or costimulatory molecules affect lymphokine secretion as well as the differentiation of Th cells? An understanding of the involvement of these interactions in the differential cytokine secretion pattern by T cells of a given specificity is crucial.

We asked whether ligands with high or low MHC-binding capabilities could result in tilting the response towards Th1 or Th2. The N-terminal nonapeptide of MBP (9.4Lys) was chosen as the test system: it has been shown that amino acid 4 is involved in MHC-binding and by altering this position 4 from Lys to Met, for example (9.4Met), the affinity of MHC interaction increases 10^4-fold.[9] B10.PL mice were immunized with several position 4 variant nonapeptides (such as 9.4 Arg, a 3-fold poorer binder than 9.4Lys, and 10 days later frequencies of Th1 or Th2 cells were assessed in draining lymph nodes using cytokine-specific, single cell, enzyme-linked immunospot (ELISA SPOT) assays. The frequency of IFN-γ producing cells was about 5 times higher in mice challenged with the highest affinity peptide, 9.4Met, in comparison with mice immunized with one of the low affinity peptides. In contrast, the frequency of IL-5 producing cells was comparable in all three groups of mice. Thus, the high affinity ligand appeared to shift the lymphokine-secretion pattern in a Th1-direction.[9] Consistent with the involvement of Th1 cells in EAE in B10.PL mice, the highest affinity ligand was able to effectively induce disease at very low concentrations (Kumar, unpublished).

Although all variants are able to induce proliferation and IL-4 secretion, only the nonamer variant with 4Met could induce significant levels of IFN-γ secretion in established type 2 CD4 T cell clones, at a relatively high peptide concentration. These data are consistent with the idea that changes in the binding affinity of a ligand to the MHC result in altered signaling owing to differences in either the display of the ligand (MHC-peptide complexes) or in the quality of the ligand on the surface of APC.[9] According to this model, the relative density of ligand on the surface of APC, and the interaction of the ligand with the TCR, will determine the cytokine secretion profile of the T cell. Furthermore, the high affinity peptide, for example 9.4Met forms relatively stable complexes with MHC molecules, resulting in high avidity interaction with the same TCR, leading to Th1 induction. These data point to a flexibility of T cell effector function that is very dependent on the physical nature of the tripartite interactions among the MHC, antigen and the TCR (see Fig. 5.1).

Recent demonstrations regarding binding of the TCR to altered peptides resulting in differential signaling (reviewed in ref. 18) are consistent with the idea that the nature of the entire stimulus brought about by various interactions, including coreceptors (e.g., CD4), costimulatory (e.g., CD28) and adhesion molecules (e.g., ICAM-1), could elicit different T cell effectors with a unique pattern of cytokine secretion. Whatever may be the mechanism of skewing the response in a type 1 or type 2 direction that results from altering the physical interaction between the TCR and the ligand, these data offer new avenues for therapy in diseases, for example, autoimmunity, allergy or AIDS, where the T cell response must be redirected into a healthy pattern.

REGULATORY T CELL CIRCUITRY AND IMMUNE DEVIATION

Regulatory CD4 T Cells Specific for a Determinant from the FR3 Region of the Vβ8.2 Chain are Spontaneously Primed in Mice Recovering from EAE

Upon induction of EAE in B10.PL or (SJL x B10.PL)F1 mouse strains, the response to the immunodominant N-terminal determinant Ac1-9 of MBP primarily utilizes the TCR Vβ8.2 gene segment.[10,19] Recent experiments have indicated that TCR-peptide specific regulatory T cells (Treg) mediate spontaneous recovery from EAE in these mice. Accordingly, CD4 Treg, directed against a V-framework determinant of the Vβ8.2 chain (B5= a.a. 76-101), become apparent during the recovery period and appear to participate in recovery, as well as in resistance to re-induction of disease.[3,20,21] Furthermore, mice vaccinated with B5 are significantly protected from antigen-induced EAE. Importantly, adoptive transfer of CD4 T cell clones specific for B5 also results in prevention of

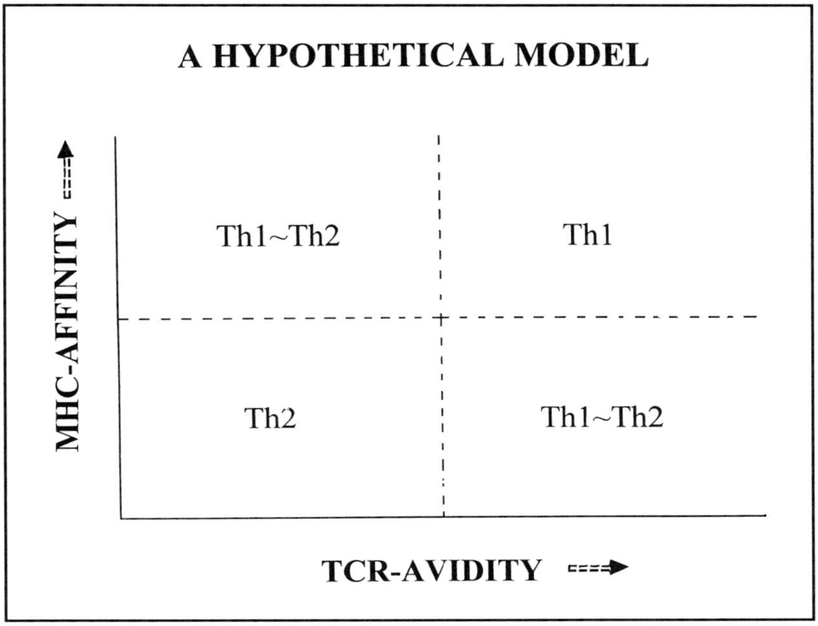

Fig. 5.1. The differentiation of antigen-specific CD4 T cell precursors is dependent on MHC-binding affinity as well as TCR avidity. In the presence of ligands with high affinity for the class II MHC molecule, Th1 skewing will occur in all those T cells in which the TCR avidity is above a certain threshold. ("TCR avidity" includes a summation of TCR affinity, binding of coreceptor CD4 molecules, engagement of costimulatory molecules (e.g., B7 family) as well as adhesion molecule involvement). Th2 skewing will occur where both MHC affinity and TCR avidity are low: in the two other quadrants of the figure, it is difficult to predict the exact extent of skewing, but a continuum of Th1/Th2 ratios is expected.

EAE. Furthermore, CD8 regulatory T cells, which recognize another Vβ8.2 determinant are also evidently recruited during this period, and the two regulatory cell types collaborate to bring about down-regulation of autoaggressive MBP-reactive T cells (ref. 3, and Kumar et al, unpublished observations).

Demonstration of the induction of TCR peptide B5-reactive CD4 T cells during the course of antigen-induced EAE suggests that a determinant within the B5 peptide from the Vβ8.2 chain of the T cell receptor is physiologically processed and presented in a class II (I-A^u) context. Determinant(s) within the B5 peptide appear to be dominant: first, lymph node cells from mice challenged with recombinant single-chain TCR molecules containing the entire Vβ8.2 chain respond to B5 in in vitro proliferative recall assays and not to other TCR peptides, although B10.PL mice can respond with vigorous proliferation to B2 (a.a. 21-50) and B4 (a.a. 61-90) (V. Kumar et al, submitted); second, irradiated spleen cells from naive animals are capable of stimulating B5-reactive, but not B4-reactive T cell clones in vitro, in the absence of exogenous peptide.[3,20,21] It appears likely that APCs constitutively process and present B5 in vivo.

Inactivation of TCR Peptide-Specific CD4 Regulatory T Cells Induces Chronic EAE

B5-specific CD4 Treg predominantly utilize TCR Vβ14 or Vβ3 gene segments. Treatment of (SJL x B10.PL)F1 mice with anti-Vβ14 and anti-Vβ3 antibodies resulted in an increase in the severity and duration of disease: even relapses were seen in one third of the Treg-depleted animals. Delayed recovery from Ac1-9-induced EAE in F1 mice following treatment with Treg-reactive-anti-Vβ monoclonal antibodies indicates that the presence of TCR-peptide-specific CD4 T cells is crucial for quick spontaneous recovery in these mice.[22] These findings delineate a physiological role for the specific TCR-peptide-reactive CD4 Treg population in the control of autoimmune disease. Oligoclonality in TCR Vβ-gene usage by spontaneously primed CD4 Treg cells was further confirmed by the demonstration that the most significant delay in recovery occurred in animals treated with both anti-Vβ14 and anti-Vβ3 mAbs. Relapses in some of the Treg-depleted animals, as well as the occurrence of chronic EAE in a mouse strain (B10.PL) that normally displays an acute episode of EAE, suggest that the spontaneous chronic autoimmune disease course found in some experimental models or in humans reflects a crucial defect in regulation.[22,23] In those individuals where MHC as well as non-MHC genes are identical, as in monozygotic twins, frequently only one suffers from chronic autoimmune disease: in such cases defective regulation could be decisive in the manifestation of the disease. It is likely that defective regulation depends upon unique encounters with microbial antigens or the presence of unique lymphocyte repertoires in each individual. Conversely, stronger and more efficient regulatory effectors may exist in EAE-resistant mouse strains, protecting them from antigen-induced autoimmunity.

How do the CD4 Treg regulate responses to MBP? It is likely that activated $V\beta 14^+$ regulatory T cells traffic to inflammatory sites in the CNS and locally downregulate Ac1-9-specific responses either through secretion of modulatory cytokines, or by recruiting regulatory CD8 T cells. Consistent with this notion, the presence of Vβ14 mRNA in CNS tissue during recovery from EAE, around day 12, has recently been demonstrated in (PL/J x SJL) mice.[24] Preliminary analysis of the cytokine secretion profile of CD4 Treg reveals that these cells are Th0-like in their ability to secrete higher amount of IL-2 and IFN-γ, as well as detectable levels of IL-4, and IL-5 (V. Kumar, unpublished observations).

Regulatory CD8 T Cells Specific for a Distinct Determinant on the Vβ8.2 Chain Also Appear to Be Involved in Regulation

CD8 deletion experiments suggest that once activated, the regulatory B5-reactive $CD4^+$ T cells recruit $CD8^+$ T cells which ultimately downregulate encephalitogenic T cells.[3,20] These CD8 cells could recognize distinct TCR determinants in an MHC class I context, displayed on the surface of MBP-specific effector cells. Recognition of the activated, encephalitogenic $V\beta8.2^+$ effector cells by TCR-peptide-specific $CD8^+$ T cells may thus be one pathway of specific regulation. Indeed, we have recently characterized a potential class I determinant within the CDR1/2 region of the Vβ8.2 chain, distinct from B5. Mice vaccinated with this peptide are completely protected from MBP/Ac1-9-induced EAE in an unusual dose-related fashion, the protective dose being very low, and higher doses being exacerbatory (Kumar et al, unpublished observations). Other experiments in mice devoid of $CD8^+$ T cells, either by anti-CD8 treatment or CD8 gene targeting, have also shown that CD8 cells play an important role in regulating EAE.[25,26,27]

Recombinant T Cell Receptor Molecules Can Prevent and Reverse EAE

Our data clearly suggest that spontaneous recovery from Ac1-9-induced monophasic EAE in B10.PL or (SJL x B10.PL)F1 mice is mediated by regulatory T cells specific for peptides derived from the Vβ8.2 chain of the T cell receptor predominantly used by the encephalitogenic T cells. Recently we have demonstrated that recombinant single chain TCRs (scTCRs) containing Vβ8.2 domains can be used as vaccines for efficient priming of both the CD4 and CD8 regulatory T cell populations. A single injection of mice with scTCRs containing Vβ8.2 domains but not containing other Vβs leads to efficient in vivo priming of Treg and significant protection from Ac1-9 or MBP-induced EAE. Notably, administration of scTCRs during ongoing disease at a 10-fold lower dose than that required for prophylactic treatment also reverses established EAE (Kumar et al, submitted). Therefore, recombinant TCRs expressing appropriate Vβ domains can be used as vaccines for the prevention and treatment of a variety of T-cell mediated pathologic conditions.

The prevention and reversal of MBP-induced EAE in F1 mice following injection with Vβ8 single chain TCR molecules (scTCRs) or single Vβ8 domains indicates that a key necessary participant in EAE is the Vβ8-expressing, Ac1-9-reactive T effector cell (Kumar et al, submitted). By selectively targeting this initial encephalitogenic T cell for down-regulation or through the preferential induction of a Th2 response, subsequent recruitment of T cells specific to other subdominant determinants of MBP may not be able to propagate disease in this system. Thus, by silencing the initiator clone inciting the inflammation, the secondary infiltrate in the target organ can be controlled. This is of direct consequence for treatment of human autoimmune conditions, where diversification following an initiating event need not imply an impossibility of successful therapy.

Dysregulation of Potentially Pathogenic MBP-Reactive T Cells Is Important for the Onset of EAE

In whatever way CD4 Treg specific for the FR3 determinant from the Vβ8.2 chain are primed in vivo, it is clear that they are activated/expanded physiologically to allow their detection during the course of EAE. There are at least two possibilities: these Treg are primed only subsequent to the onset of EAE, when the frequency of MBP-specific effector T cells reaches a threshold. Alternatively, these cells already exist as memory T cells and expand concurrently with the expansion of MBP-specific lymphocytes. Since, the presence of adjuvants boosts MBP-specific responses relatively more effectively than it enhances the regulatory repertoire, CD4 Treg are likely to depend on the generation of new FR3 sites and therefore follow the MBP-specific response, permitting the transient induction of EAE. Once the Treg expand in sufficient numbers, there can be 'spontaneous' recovery from the disease.[21,22] Recent experiments, involving mucosal exposure to Treg peptides, suggest that Treg already exist and are responsible for controlling MBP-reactive pathogenic lymphocytes in normal animals. In such animals, dysfunctional Treg cells allow an early onset and much more severe EAE (V Kumar, unpublished observations). Therefore, we favor the view that an effective balance between the Teff and Treg populations secreting particular lymphokines is crucial for the maintenance of tolerance to MBP.[23]

Immune Deviation in Treg Populations

It is probable that B5-specific regulatory CD4 T cells are at a Th0 memory state and that these cells are still susceptible to skewing towards a type 1 or type 2 response. This skewing can be of enormous significance for the outcome of the disease course and its prevention. If CD4 Treg are primed so that they predominantly secrete type 2 cytokines (for example, by nasal instillation) and thereby fail to exert their important function of recruiting CD8 Treg cells, the whole regulatory circuitry may become ineffective in controlling effector CD4 T cells; such animals should contract severe EAE with very poor recovery. Interestingly, under these conditions, indeed, MBP-specific responses become enhanced in a type 1 direction and EAE induction is augmented (Kumar et al, unpublished observations).

Therefore, it seems that a balance between type 1 and type 2 antigen-specific T cells is maintained in the first place by the structure of the antigenic determinant and in the second place, by the proper execution of the regulatory T cell circuitry.[22,23,28] Accordingly, not only is the structure of the antigenic determinant important in the causation of disease or its absence, at the level of the antigen-specific T cell, but the proper execution of the regulatory program is necessary, including both regulatory CD4 and CD8 cells, as well as the synthesis of the appropriate cytokines. It can be crucial for regulation of autoimmunity at the onset that antigen-specific CD4 T cells become skewed towards a type 2 response. However, once Th1 cells are induced, ready to cause disease, it then becomes necessary for the CD4 Treg to achieve a type 1 environment for maintenance of a disease-free state. This permits them to recruit the CD8 Treg activity and to downregulate the aggressive type 1 effectors. As an example of this, treatment with naked DNA encoding the TCR Vβ8.2 chain[29] (Kumar et al, unpublished observation), leads to regulation of EAE: we postulate that this occurs through activation of a CD4 Treg which secretes Type 1 lymphokines which then recruit CD8 to perform a coup de grace of the type 1 effectors. These initially expanding, aggressive type 1 CD4 effectors will then be specifically downregulated, leaving the Th2 cohort to take over and assist in modulating the T effector cells, thereby aborting the propagation of disease.[28] This may be a typical scenario in effective regulatory encounters.

In summary, self-antigen-specific immune deviation can be achieved either by manipulating the structure of the antigenic determinant itself as well as its flanking regions, or by the execution of TCR-based immune regulation by the members of the regulatory circuitry, fostered by the appropriate cytokine milieu. The resulting deviation can act to reestablish a healthy balance of T cells in a double-barreled way via antigen-specific and TCR-specific routes. We believe that this intricate dual aspect of cytokine regulation has many ramifications which will be unearthed in future work.

References

1. Mason D, Fowell D. T cell subsets in autoimmunity Curr Opin Immunol 1992; 4:728-732.
2. O' Garra A, Murphy K. T cell subsets in autoimmunity. Curr Opin Immunol 1993; 5:880-886.
3. Kumar V, Sercarz E. T cell regulatory circuitry: antigen-specific and TCR-idiopeptide-specific T cell interactions in EAE. Intl Rev Immunol 1992; 9:287-297.
4. Mosmann TR, Coffman RL. Th1 and Th2 cells: Different pattern of lymphokine secretion lead to different functional properties. Ann Rev Immunol 1989; 7:145-173.
5. Powell MB, Mitchell D, Lederman J et al. Lymphotoxin and tumor necrosis factor α production by MBP specific T cell clones correlate with encephalitogenicity. Intl Immunol 1990; 2:539-544.
6. Perry LL, Barzaga ME. Kinetics and specificity of T and B cell responses in relapsing experimental allergic encephalomyelitis. J Immunol 1987; 147:1434-1441.

7. Lehmann PV, Sercarz EE, Forsthuber T et al. Determinant spreading and the dynamics of the autoimmune T-cell repertoire. Immunol Today 1993; 14:203-208.
8. McRae BL, Vanderlugt CL, Dal Canto MC et al. Functional evidence for epitope spreading in the relapsing pathology of experimental autoimmune encephalomyelitis. J Exp Med 1995; 182:75-85.
9. Kumar V, Bhardwaj V, Soares L et al. MHC-ligand affinity of an antigenic determinant is crucial for differential secretion of IFN-γ or IL-4 by T cells. Proc Natl Acad Sci USA 1995; 92:9510-9514.
10. Kumar V, Kono DH, Urban JL et al. T-cell receptor repertoire and autoimmune diseases. Ann Rev Immunol 1989; 7: 657-682.
11. Acha-Orbea H, Steinman L, McDevitt HO. T cell receptors in murine autoimmune diseases. Ann Rev Immunol 1989; 7: 371-405.
12. Kotzin BL, Barr VL, Palmer E. A large deletion within the T-cell receptor beta chain gene complex in New Zealand white mice. Science 1985; 229:167.
13. Kumar V, Sercarz E. Holes in the T cell repertoire to MBP owing to the absence of the Dβ2-Jβ2 gene cluster: Implications for TCR recognition and autoimmunity. J Exp Med 1994; 179:1637-1643.
14. Kumar V, Sercarz E. Self-determinant selection and selective regulation. Chem Immunol 1994; 60:1-19.
15. Pfeiffer C, Stein J, Southwood S et al. Altered peptide ligands can control CD4 T lymphocytes differentiation in vivo. J Exp Med 1995; 181:1569-1574.
16. Segal BM, Shevach EM. IL-12 unmasks latent autoimmune disease in resistant mice. J Exp Med 1996; 184:771-775.
17. Rocken M, Saurat JH, Hauser C. A common precursor for CD4+ T cells producing IL-2 or IL-4. J Immunol 1992;148:1031-1036.
18. Kersh GJ, Allen PM. Essential flexibility in the T-cell recognition of antigen. Nature 1996; 380: 495-498.
19. Urban JL, Kumar V, Kono DH et al. Restricted use of T cell receptor V genes in murine autoimmune encephalomyelitis raises possibilities for antibody therapy. Cell 1988; 54:577-592.
20. Kumar V, Sercarz E. The involvement of TCR-peptide-specific regulatory $CD4^+$ T cells in recovery from antigen-induced autoimmune disease. J Exp Med 1993; 178:909-916.
21. Kumar V, Tabibiazar R, Geysen HM et al. The dominant self-TCR peptide from the framework III region of the β-chain controls murine EAE. J Immunol 1995; 154:1941-1950.
22. Kumar V, Stellrecht K, Sercarz E. Inactivation of TCR peptide-specific regulatory CD4 T cells leads to chronic EAE. J Exp Med 1996; in press.
23. Kumar V, Sercarz E. Dysregulation of potentially pathogenic self-reactivity is crucial for the manifestation of clinical autoimmunity. J Neurosci Res 1996; 45:334-339.

24. Bell RB, Lindsey JW, Sobel RA et al. Diverse T cell receptor V beta gene usage in the central nervous system is experimental allergic encephalomyelitis. J Immunol 1992; 150:4085-4092.
25. Jiang H, Zhang S, Pernis B, Role of $CD8^+$ T cells in experimental allergic encephalomyelitis. Science 1992; 256:1213-1215.
26. Koh D-R, Fung-Leung W-P, Ho A et al. Less mortality but more relapses in experimental allergic encephalomyelitis in $CD8^{-/-}$ mice. Science 1992; 256:1210-1213.
27. Gaur A, Haspel R, Mayer JP et al. Requirement for $CD8^+$ cells in T cell receptor peptide-induced clonal unresponsiveness. Science 1993; 259:91-94.
28. Kumar V, Sercarz E, Genetic vaccination: the advantages of going naked. Nature Med 1996; 2:857-859
29. Waisman A, Ruiz PJ, Hirschberg DL et al. Suppressive vaccination with DNA encoding a variable region gene of the T-cell receptor prevents autoimmune encephalomyelitis and activates Th2 immunity. Nature Med 1996; 2: 899-905.

CHAPTER 6

The Role of Immune Deviation in Regulation of Experimental Autoimmune Encephalomyelitis and Multiple Sclerosis

Lindsay B. Nicholson, Anja Windhagen, Mercy Prabhu Das,
David A. Hafler and Vijay K. Kuchroo

INTRODUCTION

The phenotype of the T helper immune response has a profound effect on the outcome of disease in a number of experimental models of infection[1,2] and autoimmune disease,[3] and is believed to play an important role in human diseases.[4,5] In multiple sclerosis (MS), which is postulated to be an autoimmune disease of the central nervous system (CNS), evidence suggests that the characteristic perivascular inflammation and demyelination is the result of a delayed type hypersensitivity (DTH) response to myelin protein antigens, mediated by T cells secreting Th1 cytokines (interferon-(IFN)-γ and tumor necrosis factor (TNF)-α/lymphotoxin (LT)).

In experimental autoimmune encephalomyelitis (EAE),[6,7] which can be induced by immunization with a number of different CNS derived proteins, the best studied of which are myelin basic protein (MBP)[8,9] and myelin proteolipid protein (PLP),[10-12] the immune response to these proteins has been defined in some detail. Immunization with these encephalitogenic antigens, or peptides derived from them, emulsified in complete Freund's adjuvant (CFA), leads to the generation of peptide specific $CD4^+$ Th1 cells.[8] A number of parameters influence this process and interventions which down regulate Th1 and upregulate Th2 responses have been shown to inhibit EAE.

Immunointervention in Autoimmunity by Th1/Th2 Regulation, edited by Luciano Adorini.

The understanding that such Th1 responses can be modulated in vitro by Th2 (interleukin-(IL)-4, IL-10) and Th3 (transforming growth factor-(TGF)-β) cytokines has suggested a mechanism by which a number of autoimmune and infectious processes can be regulated. This chapter will focus on factors which influence Th1/Th2 cell differentiation, experiments in EAE which show that immune deviation can protect animals from clinical disease and attempts to elucidate factors that contribute to protection. We will then review evidence that supports the hypothesis that MS is a Th1 disease and therapeutic strategies which exploit immune deviation to treat MS.

FACTORS WHICH INFLUENCE TH CELL DIFFERENTIATION

The differentiation of Th precursor (Thp) cells into Th1 and Th2 populations is an area of intense investigation and has been defined most clearly in mice. Cytokines are the most potent factors driving Th cell differentiation in vitro.[13] IL-12 induces the differentiation of Th1 cells and IL-4 the differentiation of Th2 cells. The signaling molecules STAT4 and STAT6 are known to have an important role in this process[14-17] and the proto-oncogene c-maf has been implicated as a Th2 specific transcription factor.[18] Recent studies of single cells have shown little or no evidence for coordinate expression of cytokine genes[19-22] and studies of the plasticity of Th phenotypes have shown that in vitro chronic antigen exposure leads to more stable polarization than does short term culture.[23] This suggests that T helper cell "phenotype" may be best analyzed at a population level rather than at a cellular level, and polarized phenotypes are likely to be most relevant in chronic diseases such as persistent infection and autoimmunity.

In humans polarized populations of Th cells have been generated in vitro,[24-26] and polarized responses have been identified in certain disease states[5] but the in vivo mechanisms which lead to differentiated groups of cells are uncertain. It is clear that naive T cells have the potential to differentiate along either the Th1 or Th2 pathway, and that targeting cytokines directly can effect autoimmune disease. In vivo IL-4 administration has been shown to protect animals from EAE induced with MBP specific T cells.[27,28] Conversely administering IL-12 exacerbates EAE, whilst inhibiting it with an IL-12 neutralizing antibody ameliorates disease.[29] However, in the absence of pharmacological doses of cytokine or anti-cytokine antibodies, the predominant response in vivo seems to be influenced by a number of other interactions whose relative importance has yet to be clearly delineated.

Genetic factors are known to have a significant role. In a double transgenic mouse model of diabetes it has been shown that the development of Th1 and/or Th2 responses is influenced by the genetic strain of the mouse[30] while in murine cutaneous leishmaniasis reconstitution experiments have shown that both the T cell and non-T cell compartments contribute to T helper type.[31] Susceptibility to EAE varies between mouse strains even with identical H-2 regions[32,33] and this variability may reflect in part the predominant Th type response of the strain in question. Although the genes which control this process have not been defined, recently it has been proposed that the susceptibility of BALB/c mice to infection with *Leishmania major* may relate to the inability of this mouse strain to maintain responsiveness to IL-12 after stimulation with antigen, and the genes responsible for this phenotype have been mapped to chromosome 11.[34]

Dose of antigen can play a role in T cell differentiation as shown by two in vitro studies of TCR transgenic mice, although in these studies there were differences in the role of IL-4 as judged by IL-4 neutralization studies of the primary activation of naive T cells.[35,36] In vivo in BALB/c mice low dose soluble antigen administered as a continuous infusion leads to inhibition of Th1 responses and increase in the secretion of IL-4 and IL-5.[37]

Active EAE is usually induced by antigen emulsified in CFA, which is known to predispose to the development of Th1 cells,[38] probably as a result of the induction of IL-12 by constituents of the mycobacterial cell wall and subsequent upregulation of IFN-γ. Three mechanisms which may influence T cell differentiation in EAE are currently under active investigation in our laboratories: oral feeding of myelin antigens, modulation of costimulation and the use of altered peptide ligands generated by a single or double amino acid substitution within encephalitogenic determinants.

The effect that changes in the route of antigen administration may have are exemplified in EAE by studies of oral tolerance to myelin antigens, and have been well characterized in a number of experimental systems.[39] Several mechanisms which lead to peripheral tolerance and are dependent upon the dose and frequency of fed antigen have been defined. At high antigen concentrations deletion by apoptosis and induction of anergy may be the predominant tolerizing mechanisms,[40] but at lower antigen doses it has been possible to demonstrate the induction of T cells which secrete the Th2 cytokines IL-4 and IL-10 and also Th3 cells which secrete TGF-β.[41,42] Both these cell types can protect SJL mice by bystander mechanisms from the induction of EAE by MBP and PLP.

Work from our group[43] has also shown that modulating the costimulatory molecules B7-1 (CD80) and B7-2 (CD86) has profound effects on EAE by affecting the differentiation of antigen specific T cells. Treating SJL mice with anti-B7-1 monoclonal antibody (mAb), following immunization with the encephalitogenic peptide PLP 139-151, reduces the severity of disease. Conversely treatment with anti-B7-2 mAb exacerbates disease. T cell differentiation was shown to play a key role in this process by demonstrating changes in the levels of IFN-γ measured following the stimulation of lymph node cells derived from mice in the various treatment groups. IFN-γ was decreased in the mice treated with anti-B7-1 and increased in the mice treated with anti-B7-2 and the effects of anti-B7-1 antibody could be reversed by the administration of anti-IL-4. Furthermore, antigen specific T cell clones derived from mice treated with control immunoglobulin were predominantly (87%) of a Th1 phenotype whilst those derived from mice treated with anti-B7-1 were mainly (80%) Th2. These findings have been confirmed in a number of different EAE models using different myelin antigens to induce disease.[44] Work using anti B7-1 F(ab) fragment therapy to block relapsing EAE confirmed that blocking B7-1 costimulation inhibits EAE.[45] However, in other systems the use of similar antibodies has different results and anti B7-2 treatment is more effective than anti B7-1 treatment in inhibiting diabetes in the NOD mouse.[46] This may relate to the phase of the disease process in which the antibodies are administered.

In EAE induced with PLP 139-151, changing single amino acids in the encephalitogenic peptide generates altered peptides which protect from disease induced with the native peptide.[47] We have shown that some of these analogues protect by inducing antigen specific cells which are cross-reactive with the native peptide but secrete Th2/Th0 rather than Th1 cytokines on activation.[48] This change in phenotype occurs despite immunizations being carried out in the presence of CFA. A study of the effects of altered MBP analogues upon the cytokine profile of lymph node cells obtained from rats following primary immunization reported reduced production of IFN-γ and TNF-α; however, there was no analysis of Th2 cytokine production.[49] A role for antigen in the differentiation of T cells has also been described in an experimental system using peptides derived from human collagen IV. In this system immunizing H-2^b mice with wild type peptide or mutants of it led to differential expression of IFN-γ and IL-4 mRNA by CD4 T cells stimulated in primary lymph node cell culture.[50]

CHARACTERISTICS OF AUTOPATHOGENIC T CELLS

Immunization with encephalitogenic antigens leads to the generation of $CD4^+$ T cells which can induce disease in naive animals following adoptive transfer. However, specific recognition of myelin antigen is not sufficient to ensure the development of disease as shown by the low frequency of spontaneous EAE (14%) in transgenic mice expressing a T cell receptor (TCR) specific for the MBP epitope Ac1-9.[51] Since autopathogenic T cell clones derived from mice immunized with encephalitogenic MBP or PLP peptides are found to be of the Th1 phenotype[8,10] when analyzed on the basis of IFN-γ/IL-2 versus IL-4/IL-10 production, the role of individual Th1 cytokines has been analyzed by several different approaches.

The elaboration of a full range of Th1 cytokines is not necessary for the development of EAE, since mice with a targeted disruption of the IFN-γ gene are susceptible to disease,[52] although interfering with the generation of Th1 cells by blocking IL-12 does inhibit the development of EAE.[29] TNF-α and LT/TNF-β, which are known to kill oligodendrocytes in vitro, have been correlated with disease by Ruddle et al[53] who reported that the ability of T cell clones to transfer EAE was proportional to the detectable levels of TNF-β mRNA. In addition it was shown that a single dose of TNF-β neutralizing antibody blocked the development of disease induced by adoptive transfer of encephalitogenic cells. Chung et al[54] have shown that the characteristics of TNF-α production by LPS or IFN-γ stimulated astrocytes, derived from strains of rats susceptible or resistant to EAE, was different between these strains, and inferred that this may relate to clinical disease.

A further important property of encephalitogenic T cells is the expression of appropriate cell surface addressins which allow trafficking to sites of inflammation within the CNS. The importance of the α4β1 integrin VLA-4 has been shown in a number of studies. EAE induced by adoptive transfer can be inhibited by the administration of anti-α4 integrin mAb.[55] Baron et al[9] reported data from two daughter clones derived from an encephalitogenic parent and showed that their encephalitogenic potential related to the surface expression of VLA-4, but not to the levels of IFN-γ or TNF-α/β. In EAE caused by the transfer of PLP specific clones, clones lacking the surface expression of VLA-4 were only able to induce disease following the administration of intravenous heat-killed *Bordetella pertussis*, a procedure which is thought to increase the permeability of the blood-brain barrier.[56] These data suggest that the ability of autopathogenic T cells to enter the CNS is critical to the development of clinical disease.

In active inflammatory CNS diseases such as MS, it has been shown that lymphocytes isolated from the CSF adhere to vascular endothelial cells 10 times more readily than peripheral blood lymphocytes from the same patients. This increased adherence was not CNS specific since it occurred on both CNS and non-CNS derived endothelium.[57] In immunocytochemical studies of lesions from patients with potentially MS relevant adhesion molecules have been identified[58] but none have been specifically associated with this disease.[59]

There is convincing evidence that Th1 type cytokines are localized to inflammatory lesions in the brains of animals with EAE. In a rat model of this disease immunohistochemical staining showed that at the peak of disease perivascular infiltrates within the CNS stained positive for the inflammatory cytokines IL-1, -2, -6 -8, TNF-α and IFN-γ. Recovery was associated with the appearance of TGF-β and IL-4, cytokines which would be expected to suppress ongoing inflammation, whilst tolerance induced by oral feeding was accompanied by earlier detection of these anti-inflammatory cytokines.[60] Kennedy et al[61] used semi-quantitative reverse transcriptase-polymerase chain reaction (RT-PCR) to measure cytokine mRNA in the spinal cord of mice with EAE. They showed that IL-2, -4, -6 and IFN-γ peak approximately together whilst

IL-10 peaks later during the recovery stage. Using a quantitative PCR technique in a study of specific CNS cell types in SJL mice, Renno et al[62] found that TNF-α was transcribed predominantly from microglia and macrophages while, as expected, IL-2 and IFN-γ were produced by infiltrating $CD4^+$ cells.

THE ROLE OF TH2 CELLS

The clinical course of EAE varies with the specific animal model used, and also with the protocol used to induce the disease. In most models the acute disease is followed by a remission which may be permanent and accompanied by resistance to further disease induction (e.g., in Lewis rats[39]) or may be transient and followed by subsequent relapses and remissions.[63] Relapse of disease has been linked to determinant spreading,[64,65] but mechanism(s) which mediate remissions and recovery are less well characterized.

The presence of Th2 cytokines within the brains of animals recovering from EAE has led to the hypothesis that remission of disease might be related to the presence of antigen specific T cells secreting Th2 (IL-4 and IL-10) and Th3 (TGF-β) cytokines. It is known that Th2 cells do not induce EAE,[9] but direct evidence that $CD4^+$ Th2 and Th3 cells can ameliorate EAE required adoptive transfer of protective T cell clones. Chen et al[42] exploited the oral tolerance model and derived $CD4^+$ T cell clones from SJL mice fed and immunized with MBP. The majority (42/48) of these mucosally derived MBP specific T cell clones secreted TGF-β at either high or lower levels but it was also possible to generate MBP specific clones secreting predominantly IL-4 and IL-10. Both a representative TGF-β secreting clone and an IL-4/IL-10 secreting clone protected actively immunized animals from EAE. The TCR Vα and Vβ regions of the Th3 clone were found to be identical to that of a pathogenic Th1 clone.

Th2 clones specific for PLP 139-151 that secrete high levels of IL-4 and IL-10 have also been derived by immunization with the encephalitogenic peptide in the presence of anti-B7-1[43] and shown to inhibit the induction of EAE by transfer at the time of active immunization, or at the onset of clinical disease. In the same disease model we have shown that immunizing mice with Q144, an analogue of the PLP 139-151 peptide, results in the expansion of T cells which are predominantly of a Th0/Th2 phenotype. Transfer of T cell lines specific for the altered peptide protect animals from EAE.[48] Using a second APL of PLP 139-151 (L144/R147),[47] which in vitro is a T cell receptor antagonist for Th1 cells, also generates Th2 cells in vivo and mediates bystander suppression of EAE caused by unrelated myelin encephalitogens including MBP and myelin oligodendrocyte glycoprotein.[66]

Targeting antigen to B cells using covalently linked MBP-IgD results in the generation of a Th2 response which prevents EAE in rats.[67,68] Soluble peptide antigen (or a peptide analog) administered i.p. can also reverse EAE induced by transfer of an MBP specific encephalitogenic T cell clone, in a manner which is IL-4 dependent.[69] In this system the mechanism appears to be analogous to high dose tolerance induced by aqueous protein antigens.[70,71] Both these groups found a reduction in infiltration of leukocytes into the brains of protected animals. This was thought to be due either to a failure of migration into the brain in the case of the rats or to a specific efflux of pathogenic cells from the brain following treatment with peptide.

In contrast to these studies, other groups have established antigen specific Th2 lines or clones which are not protective in vivo. Khoruts et al used exogenous IL-4 to differentiate T cell lines towards a Th2 phenotype and found that lines with this phenotype could not suppress disease induced by Th1 cells.[72] These results are concordant with experiments in the NOD mouse where diabetes antigen specific transgenic T cells caused disease when differentiated to a Th1 phenotype in vitro, but did not mediate protection when differentiated to a Th2 phenotype.[73]

CHARACTERISTICS OF "PROTECTIVE" TH2 CLONES

Since the data shows that antigen specific Th2 cells can influence the course of clinical disease in EAE but do not necessarily do so, determining the factors which distinguish cells of a protective phenotype is an important goal. A number of factors may contribute to effective regulation of DTH like immune responses, and they may exert their major influence at induction of the immune response, in the effector phase or during both.

The effects of exogenous IL-4 on EAE vary with the time at which it is administered. In disease induced by transfer of encephalitogenic MBP specific T cells, protection induced by IL-4 is greatest when it is given for 11 days from the time of transfer. Later administration (days 6-11) was much less effective in protecting animals from disease.[27] In the effector phase IL-10 and TGF-β may have a greater anti-inflammatory effect through their actions on antigen presenting cells as well as T cells. However, some Th2 cytokines may exacerbate autoimmune disease when expressed at high levels. Constitutive expression of IL-10 in the pancreas of transgenic NOD mice leads to accelerated spontaneous diabetes,[74] which in the context of the NOD MHC is sufficient to ensure that disease develops.[75] In contrast transgenic mice overexpressing IL-10 in the T cells are protected from the development of EAE (MPD and VKK unpublished data).

To allow traffic to inflammatory lesions within the CNS, regulatory cells must express appropriate cell surface addressins, since if appropriate localization does not occur, Th2 and Th3 cells will not receive the antigen specific signals necessary for activation and cytokine release. Also in the effector phase of the disease Th2/Th3 regulatory cells present within lesions must compete effectively with autoaggressive cells for antigen on APCs. In this regard we have recently completed an analysis of the recognition of PLP 139-151 by conventionally derived Th1 cells compared with Th2 cells derived in the presence of anti-B7-1. This showed that following immunization with PLP 139-151 in CFA the T cell immune response comprises a population of Th1 cells expressing a diverse repertoire of T cell receptors, but which depends on recognition of the same central amino acid, tryptophan 144, as the primary TCR contact residue.[47] In contrast, analysis of protective Th2 T cell clones shows them to be drawn from a population of cells which focus towards the N-terminal end of the peptide with little requirement for tryptophan 144. We hypothesize that Th2 cells which have shifted their recognition motif may compete more effectively with Th1 cells for the activating ligand than Th2 cells which share the same motif, perhaps because of differences in the avidity of interaction with MHC/peptide. APL specific Th2 clones, which are also likely to recognize a different motif, may be effective in regulating EAE for the same reason. If this is the case then therapy with altered peptides, or proteins administered via alternative routes such as orally (leading to the generation of cross-reactive Th2 cells and/or cells with a Th2 recognition motif), might be more successful than therapy with cytokines which cause immune deviation of Th1 cells. Factors which are believed to be important in the development of autopathogenic and autoregulatory cells are summarized in Table 6.1.

CYTOKINES IN THE CEREBROSPINAL FLUID (CSF)

Studies in MS to determine how closely the human disease conforms to the Th1/Th2 paradigm have provided some experimental support for the relevance of this model. Several groups have investigated the Th1 cytokines TNF-α and IFN-γ in the cerebrospinal fluid in MS. Sharief and Hentges[76] reported high levels of TNF-α in the CSF of 53% of patients with chronic progressive MS as compared to none of the patients with stable disease and also correlated TNF-α levels with the severity and progression of the disease. Similar findings were reported by Tsukada et al.[77] Maimone et al[78] found TNF-α levels of up to 140 pg/ml in the CSF of 23% of MS patients and 29% of patients with other inflammatory neurological diseases using an ELISA with a sensitivity of 40 pg/ml. In studies of other neurologic conditions, TNF-α was found to be elevated in the serum and CSF of patients with Guillain-Barré syndrome (which may also have an autoimmune etiology) and to correlate with disease severity, but similar levels of cytokine were also found in patients with polyneuropathies and other neurological diseases when compared to healthy controls.[79] Elevated levels of TNF-α have also been detected in the CSF of patients with HIV-related neurologic disorders.[80] In contrast to these findings other investigators do not find differences in TNF-α and IFN-γ levels between MS and control subjects.[81-85]

Table 6.1. Factors which may influence the development of autopathogenic and autoregulatory cells

	Autoaggressive T cells	Autoregulatory T cells
Induction Phase		
Cytokines in microenvironment	IL-12[29]	IL-4[27,28]
Route	subcutaneous in CFA[38]	oral[39,41,42] targeted to B cells[48,49]
Structure	native antigen in emulsion particulate antigen	soluble antigen[37,69] altered antigen[48,49]
Costimulation	increased in presence of anti-B7-2[43,44]	increased in presence of anti-B7-1[43,44]
Effector Phase		
Cytokines secreted by T cells	TNF-α/β, IFN-γ[8,10,53,60]	IL-4, IL-10[41-43,60]
Trafficking	inhibited by anti-VLA-4[9,55]	unknown
Recognition motif	recognizes primary TCR contact	shifted from primary TCR contact

This suggests that increased levels of cytokine in the circulation is not specific for demyelinating disease but correlates better with disease activity.

CYTOKINES FROM T CELLS IN MULTIPLE SCLEROSIS

Studies of T cells are likely to be more relevant to defining patterns of cytokine secretion in MS, since these cells may be initiating or maintaining antigen specific responses. Using 2-3 week old T cell lines generated in the presence of recombinant IL-2 from the peripheral blood of MS patients with active disease, Mokhtarian et al[86] reported increased production of proinflammatory cytokines (IL-2, IFN-γ, TNF-α /TNF-β) compared to lines obtained from patients with other autoimmune diseases and to normal controls. TGF-β production, mainly from $CD8^+$ and $CD45RA^+$ T cells, was reduced in the MS patients and this reduction was correlated with disease activity. Using PCR to measure cytokine mRNA, Rieckmann et al examined peripheral blood mononuclear cells isolated from MS patients and found elevated levels of TNF-α.[87] A longitudinal study of patients with relapsing remitting MS, reported by the same group, found that increased expression of TNF-α preceded relapses by 4 weeks and at this time IL-10 and TGF-β were decreased. No correlation with disease activity was observed for IL-1, IL-4, IL-6 or IFN-γ.[88] Zipp et al[89] have recently compared CD4+ T cell lines from HLA-DR2 positive and DR2 negative MS patients and normal controls to evaluate the effect of this MS associated haplotype on TNF/LT production. TNF-α and LT secretion was higher in T cell lines from all DR2 positive subjects, possibly due to the genetic linkage of the DR and the TNF gene locus on chromosome 6. No difference was found in the levels of IFN-γ production. Comparing PLP reactive T cell lines generated from patients during acute relapses of MS and during remission, it was found that during relapses lines produced IFN-γ and TNF-α/β but no TGF-β. During remission T cell lines from the same patients produced IL-10 and TGF-β more frequently.[90]

We have previously described T cells cloned directly from the peripheral blood and CSF of MS patients, noninflammatory disease patients or normal controls. Most of these T cell clones showed a Th0 cytokine profile, and no Th2 clones were detected.[91] Importantly, a Th1 subset occurred in up to 74% of the MS patients but no Th1 clones were detected in noninflammatory controls. This observation is supported by a study which found increased numbers of IFN-γ secreting T cells in the CSF of MS patients. These cells produced IFN-γ upon activation by several myelin antigens and several MBP peptides.[92]

Romagnani and co-workers have shown that cytokine secretion in human T cells may be regulated by IL-4 and IL-12.[24,25] In vitro culture of mononuclear cells with IL-12/anti-IL-4 mAb induces IFN-γ secretion while IL-4/anti-IL-12 induces IL-4 (Th2). We recently showed that IL-12 in humans induces a T cell subset characterized by simultaneous secretion of IFN-γ and IL-10.[26] Using a similar system to compare MS patients and normal controls we did not observe differences in cytokine secretion profiles between the two groups (Windhagen et al, submitted for publication). This demonstrates that secretion of cytokines can be altered in MS patients as well as in normal subjects, indicating that even at the later stages of the disease immune deviation may be effective in switching the function of myelin reactive T cells.

Much less is known about the role of CD8+ T cells in human autoimmune disease. In EAE regulatory CD8+ T cells have been described and shown to mediate suppression by secretion of TGF-β or IL-10.[41,93] Antigen nonspecific abnormalities in T cell mediated suppression have been described in the chronic progressive form of MS.[94-97] Using the autologous mixed lymphocyte reaction it has been shown that CD8+ T cells mediate suppression by production of IFN-γ and that this suppression is defective in chronic progressive MS.[98]

CYTOKINES IN BRAIN TISSUE

In active chronic MS lesions IFN-γ and HLA-DR were associated with astrocytes at the edges of lesions and in the normal appearing white matter where CD4+/IL-2 receptor positive T cells were found.[99] Hofman et al[100] described positive staining for IL-1, IL-2 and IL-2R in MS lesions and detected TNF-α associated with astrocytes in 6 MS brains. TNF-α was also detectable in lesions from subacute sclerosing pan-encephalitis (SSPE) brain but not in normal or Alzheimer's disease brains. Selmaj et al[101] identified TNF-α and TNF-β using immunocytochemistry in acute and chronic MS lesions but not in spleen or PBMCs from MS patients. TNF-α was associated with astrocytes and macrophages. TNF-β expression was found in association with CD3+ cells and microglia and high levels of TNF-β were also found in adrenoleukodystrophy. These authors also identified TNF-α and TNF-β in other inflammatory neurological diseases but not in Alzheimer's and Parkinson's disease. In studies of other diseases, elevated levels of TNF-α and decreased levels of IL-4 mRNA expression have been described in HIV-associated dementia.[102] In postmortem brains of SSPE patients with extensive cellular infiltration and demyelination, Nagano et al[103] found IL-1, IL-2, IL-6, TNF-α and LT using immunocytochemistry.

We recently investigated the expression of cytokines and costimulatory molecules in a panel of histologically characterized CNS specimens from MS and control cases using semiquantitative RT-PCR and immunocytochemistry.[104] The costimulatory molecules B7-1, B7-2 and the cytokine IL-12p40 were upregulated in acute MS plaques from early disease cases, as indicated by increased levels of mRNA and B7-1 protein. B7-1 staining was localized to the perivenular inflammatory cuffs but not the parenchyma. The expression of these molecules was compared in the brain tissue of a single subject with both MS and a cerebral infarct and increased expression of B7-1 and IL-12p40 was found in the MS plaques, but not in the infarct area. The differences in cytokine mRNA expression were specific for IL-12p40 and no differences were found for IL-2, IL-4, IL-6, IL-13, TNF-α, TNF-β, IFN-γ and IL-12p35. These findings suggest that an early event in the initiation of MS involves upregulation of costimulatory molecules and IL-12, resulting in conditions that maximally stimulate T-cell activation and Th1 type immune responses. Both we and others[102] have noted that T cell derived cytokines (IL-2, IL-4, IL-10 and IFN-γ) are difficult to detect even using a very sensitive radioactive PCR method.

Woodroofe and Cuzner[105] used in situ hybridization to detect cytokine mRNA in MS brains. mRNAs for all cytokines under investigation (IL-1β, IL-2, IL-4, IL-6, IL-10, TNF-α, IFN-γ, TGF-β1/2) were detected in MS brains. In inflammatory demyelinating lesions TNF-α, IFN-γ and IL-6 were predominant, whereas inflammatory nondemyelinating lesions showed a relative predominance of IFN-γ and IL-2. All cytokines were expressed in perivascular infiltrates but not the parenchymal cells suggesting they come mainly from circulating cells that cross the blood brain barrier.

Studies of brain tissue obtained from patients in the late stages of AIDS, demonstrating the proinflammatory cytokines IL-1, IL-6, IFN-γ and TNF-α mainly in endothelium and microglia,[106] and the description of increased levels of TNF-α in other neurologic diseases like GBS and SSPE, raises the possibility that these cytokines might be a nonspecific marker of immune activation as well as a critical event in the initiation of MS. However, the evidence described above supports a model in which early inflammatory lesions in MS brain are a pro-inflammatory, pro-Th1 milieu from which pathogenic T cells that orchestrate demyelination emerge. Manipulations that shift the balance in favor of anti-inflammatory Th2 cells have been successful in treating experimental autoimmune diseases, and such strategies may be effective in treating MS (Fig. 6.1).

IMMUNE MODULATION IN HUMAN AUTOIMMUNE DISEASE

A number of strategies which may induce immune deviation in organ-specific autoimmune disease have been proposed. Direct administration of IL-4 is an effective treatment of EAE, and although the use of high doses of IL-4 has had significant adverse effects in the treatment of human tumors, it has been suggested that at lower doses it may be a useful therapy.[28] Administration of high doses of soluble autoantigen also favors the induction of Th2 cells,[69-71] but this seems unlikely to be adopted as a therapy for the treatment of human disease.

Cyclophosphamide is used to treat progressive multiple sclerosis, as well as other aggressive immune mediated diseases. Work from our center has recently shown that $CD4^+$ T cells from patients treated with a combination methylprednisolone and cyclophosphamide release IL-4 when activated with anti-CD3. The amount of IL-4 secreted is 10-fold greater than that secreted by untreated patients ($p = 0.001$) and significantly greater than patients treated with methylprednisolone alone. There was also a positive correlation between IL-4 secretion and eosinophilia.[107] This shows that immune deviation can be induced by nonantigen specific immunosuppressive treatment.

A number of other drugs are potential candidates for therapy intended to cause immune deviation in patients with MS. The phosphodiesterase inhibitor pentoxifylline has been tested for this effect in patients with MS and appears to favor the development of Th2 responses.[108] In mice, Schiff-base-forming-drugs enhance T cell responses and selectively favor the development of Th1 responses.[109] It is therefore possible that manipulation of these compounds may also be able to induce Th2 responses.

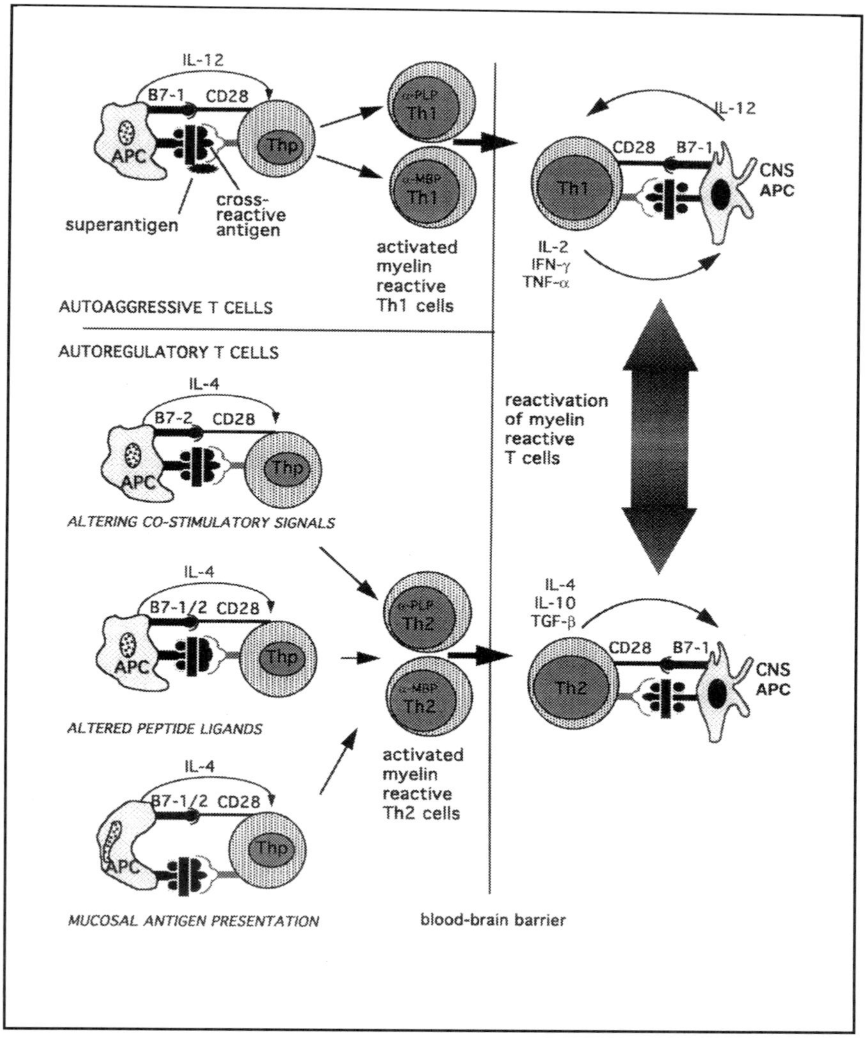

Fig. 6.1. Myelin antigen reactive T cells, which are activated and develop into Th1 cells that can traffic to the CNS, cause inflammation and demyelination. Under different activating conditions (altered costimulatory signals, altered peptide ligands, mucosal presentation) antigen-specific cells of a Th2/Th3 phenotype develop and these can regulate the pro-inflammatory action of Th1 cells. (Adapted with permission from Windhagen et al, Chemical Immunology 63, 171-186 1996. S Romagnani (Ed), Karger, Basel.)

IFN-β is an effective therapy for relapsing-remitting-MS. T lymphocytes from patients treated with IFN-β and stimulated with anti-CD3 before and after treatment showed significant decreases in TNF-α and increases in IL-6 (a Th2 cytokine).[110] Although this treatment has multiple effects on the immune system[111] these results are consistent with one of these being immune deviation of T cells. In contrast treatment with IFN-γ exacerbates multiple sclerosis, leading to an increased rate of relapse.[112,113]

The generation of antigen specific Th2 and Th3 cells is also one of the mechanisms by which low dose oral tolerance prevents EAE. Oral administration of myelin[114] and collagen[115] is well tolerated in studies of patients with multiple sclerosis and rheumatoid arthritis, and clinical trials to assess the therapeutic benefit of these treatments are underway. To define further whether oral administration of myelin antigens induced antigen specific cells of a Th2 or Th3 phenotype, we examined patients who had been treated with oral myelin for at least 2 years in comparison with nontreated patients. Treated patients were shown to have an increase in the relative frequency of MBP and PLP specific TGF-β1-secreting T cells lines.[116] Cells of this type have the potential to regulate Th1 DTH -like immune responses at sites of inflammation within the CNS in EAE and demonstrating their presence following oral feeding of myelin supports the notion that similar mechanisms of tolerance occur in patients.

CONCLUSION

Advances in basic understanding of the phenotype of $CD4^+$ T cells in mice and humans has led to their division into Th1 and Th2 cells. These cell types appear to play a critical role in the induction and regulation of experimental and human autoimmune disease. Because these functionally distinct populations cross-regulate each other in vitro it was hypothesized that similar effects would be seen in vivo and this has been shown to be the case. However, experiments in vivo make it clear that the function of T cells, defined in terms of cytokine production, is not sufficient to determine whether a myelin antigen specific Th1 cell will be auto-aggressive, or whether a Th2 cell with the same specificity will be protective. The differences between Th2 cells are highlighted by data which shows that the methods that are used to generate Th2 cells have a profound influence on whether such cells protect against EAE, and diabetes in the NOD mouse.

Some progress has been made in defining which parameters lead to auto-aggressive Th1 cells. Much less is known about the factors that define "protective" Th2 cells. We have shown that protective Th2 cells have different fine specificity than autoaggressive Th1 cells despite being activated by the same peptide. This may have an influence upon their ability to compete with Th1 cells in a micro-environment where a limited amount of antigen is being presented. The ability of regulatory cells to traffic to areas of inflammation is clearly another important property which may be relevant in preventing disease and there are a number of other possibilities which have not yet been addressed experimentally.

In MS there is accumulating evidence that the disease is mediated at least in part through pro-inflammatory Th1 cytokines. It appears that in acute MS lesions the micro-environment favors the generation of Th1 T cells. However, the progression from these early lesions to demyelination is not yet fully understood.

Because not all routes to immune deviation have the same outcome in terms of protection, not all therapeutic strategies which cause immune deviation will have equal benefit. Based on the experimental evidence above, regimes which lead to the generation of separate populations of regulatory Th2 cells may be more effective than strategies which change the phenotype of cells which would otherwise be autoaggressive. This may, in part, underlie the effectiveness of oral tolerance and altered peptide ligand therapy in EAE. Recent evidence in patients has shown that a spectrum of different therapies can lead to immune deviation. This is encouraging and clearly justifies the ongoing clinical work in this field.

Acknowledgments

Work in the authors' laboratories is supported by the NIH (R01NS24247, R01N535685 and R29 NS-30843) and the National Multiple Sclerosis Society (RG2571A2 and RG2320B3). LBN is a Postdoctoral Fellow of the National Multiple Sclerosis Society.

References

1. Mosmann TR, Coffman RL. Th1 and Th2 cells: Different patterns of lymphokine secretion lead to different functional properties. Annu Rev Immunol 1989; 7:145-73.
2. Sher A, Coffman RL. Regulation of immunity to parasites and T cell-derived cytokines. Annu Rev Immunol 1992; 10:385-409.
3. Mosmann TR, Sad S. The expanding universe of T-cell subsets: Th1, Th2 and more. Immunol Today 1996; 17:138-46.
4. Powrie F, Coffman RL. Cytokine regulation of T-cell function: potential for therapeutic intervention. Immunol Today 1993; 14:270-4.
5. Romagnani S. Lymphokine production by human T cells in disease states. Annu Rev Immunol 1994; 12:227-57.
6. Wekerle H. Experimental autoimmune encephalomyelitis as a model of immune-mediated CNS disease. Current Opinion in Neurobiology 1993; 3:779-84.
7. Alvord EC, Kies MW, Suckling AJ. Experimental allergic encephalomyelitis: a useful model for multiple sclerosis. New York: Liss; 1984.
8. Zamvil SS, Steinman L. The T lymphocyte in experimental allergic encephalomyelitis. Annu Rev Immunol 1990; 8:579-621.

9. Baron JL, Madri JA, Ruddle NH et al. Surface expression of α4 integrin by CD4 T cells is required for their entry into brain parenchyma. J Exp Med 1993; 177:57-68.
10. Kuchroo VK, Sobel RA, Laning JC et al. Experimental allergic encephalomyelitis mediated by cloned T cells specific for a synthetic peptide of myelin proteolipid protein. J Immunol 1992; 148:3776-82.
11. Sobel RA, Greer JA, Kuchroo VK. Autoimmune responses to myelin proteolipid protein. Neurochem Res 1994; 19:915-21.
12. Greer JM, Sobel RA, Sette A et al. Immunogenic and encephalitogenic clusters of myelin proteolipid protein. J Immunol 1996; 156:371-9.
13. Paul WE, Seder RA. Lymphocyte responses and cytokines. Cell 1994; 76:241-51.
14. Thierfelder WE, van Deursen JM, Yamamoto K et al. Requirement for Stat4 in interleukin-12-mediated response of natural killer and T cells. Nature 1996; 382:171-4.
15. Kaplan MH, Sun Y, Hoey T et al. Impaired IL-12 responses and enhanced development of Th2 cells in Stat4-deficient mice. Nature 1996; 382:174-7.
16. Takeda K, Tanaka T, Shi W et al. Essential role of Stat6 in IL-4 signalling. Nature 1996; 380:627-30.
17. Shimoda K, Vandeursen J, Sangster MY et al. Lack of IL-4 induced Th2 response and IgE class switching in mice with disrupted Stat6 gene. Nature 1996; 380:630-3.
18. Ho I, Hodge MR, Rooney JW et al. The proto-oncogene c-maf is responsible for tissue-specific expression of interleukin-4. Cell 1996; 85:973-83.
19. Kelso A. Th1 and Th2 subsets: paradigms lost? Immunol Today 1996; 16:374-9.
20. Bucy RP, Panoskaltsis-Mortari A, Huang G et al. Heterogeneity of single cell cytokine gene expression in clonal T cell populations. J Exp Med 1994; 180:1251-62.
21. Bucy RP, Karr RW, Huang G et al. Single cell analysis of cytokine gene expression during $CD4^+$ T-cell phenotype development. Proc Natl Acad Sci USA 1995; 92:7565-9.
22. Openshaw P, Murphy EE, Hosken NA et al. Heterogeneity of intracellular cytokine synthesis at the single-cell level in polarized T helper 1 and T helper 2 populations. J Exp Med 1995; 182:1357-67.
23. Murphy E, Shibuya K, Hosken N et al. Reversibility of T helper 1 and 2 populations is lost after long-term stimulation. J Exp Med 1996; 183:901-13.
24. Maggi E, Parronchi P, Manetti R et al. Reciprocal regulatory effects of IFN-γ and IL-4 on the in vitro development of human Th1 and Th2 clones. J Immunol 1992; 148:2142-7.
25. Manetti R, Parronchi P, Giudizi MG et al. Natural killer cell stimulatory factor (interleukin 12 [IL-12]) induces T helper type 1 (Th1)-spe-

cific immune responses and inhibits the development of IL-4-producing Th cells. J Exp Med 1993; 177:1199-204.
26. Windhagen A, Anderson DE, Carrizosa A et al. IL-12 induces human T cells secreting IL-10 with IFN-γ. J Immunol 1996; (157):1127-31.
27. Racke MK, Bonomo A, Scott DE et al. Cytokine-induced immune deviation as a therapy for inflammatory autoimmune disease. J Exp Med 1994; 180:1961-6.
28. Rocken M, Racke MK, Shevach EM. IL-4 induced immune deviation as antigen-specific therapy for inflammatory autoimmune disease. Immunol Today 1996; 17:225-31.
29. Leonard JP, Waldburger KE, Goldman SJ. Prevention of experimental autoimmune encephalomyelitis by antibodies against interleukin 12. J Exp Med 1995; 181:381-6.
30. Scott B, Liblau R, Degermann S et al. A role for non-MHC genetic polymorphism in susceptibility to spontaneous autoimmunity. Immunity 1994; 1:73-82.
31. Shankar AH, Titus RG. T cell and non-T cell compartments can independently determine resistance to Leishmania major. J Exp Med 1995; 181:845-55.
32. Kalman B, Knobler RL, Lublin FD. T cell receptor V gene utilization in myelin basic protein specific clones from CXJ1 recombinant inbred mice. J Neuroimmunol 1993; 43:191-4.
33. Encinas JA, Weiner HL, Kuchroo VK. Inheritance of susceptibility to experimental autoimmune encephalomyelitis. [In Press] Journal of Neuroscience Research 1996
34. Guler ML, Gorham JD, Hsieh C et al. Genetic susceptibility to Leishmania: IL-12 responsiveness in Th1 cell development. Science 1996; 271:984-7.
35. Constant S, Pfeiffer C, Woodard A et al. Extent of T cell receptor ligation can determine the functional differentiation of naive $CD4^+$ T cells. J Exp Med 1995; 182:1591-6.
36. Hosken NA, Shibuya K, Heath AW et al. The effect of antigen dose on $CD4^+$ T helper cell phenotype development in a T cell receptor-αβ-transgenic model. J Exp Med 1995; 182:1579-84.
37. Guery JC, Galbiati F, Smiroldo S et al. Selective development of T helper (Th)2 cells induced by continuous administration of low dose soluble proteins to normal and 2 microglobulin-deficient BALB/c mice. J Exp Med 1996; 183:485-97.
38. Janeway CAJ, Carding S, Jones B et al. $CD4^+$ T cells: Specificity and function. Immunol Rev 1988; 101:39-80.
39. Weiner HL, Friedman A, Miller A et al. Oral tolerance: Immunologic mechanisms and treatment of animal and human organ-specific autoimmune diseases by oral administration of autoantigens. Annu Rev Immunol 1994; 12:809-37.

40. Chen Y, Inobe J, Marks R et al. Peripheral deletion of antigen-reactive T cells in oral tolerance. Nature 1995; 376:177-80.
41. Miller A, Lider O, Roberts AB et al. Suppressor T cells generated by oral tolerization to myelin basic protein suppress both in vitro and in vivo immune responses by the release of transforming growth factor β after antigen-specific triggering. Proc Natl Acad Sci USA 1992; 89:421-5.
42. Chen Y, Kuchroo VK, Inobe J et al. Regulatory T cell clones induced by oral tolerance: Suppression of autoimmune encephalomyelitis. Science 1994; 265:1237-40.
43. Kuchroo VK, Prabhu Das M, Brown JA et al. B7-1 and B7-2 costimulatory molecules activate differentially the Th1/Th2 developmental pathways: application to autoimmune disease therapy. Cell 1995; 80:707-18.
44. Racke MK, Scott DE, Quigley L et al. Distinct roles for B7-1 (CD-80) and B7-2 (CD-86) in the initiation of experimental allergic encephalomyelitis. J Clin Invest 1995; 96:2195-203.
45. Miller SD, Vanderlugt CL, Lenschow DJ et al. Blockade of CD28/B7-1 interaction prevents epitope spreading and clinical relapse in murine EAE. Immunity 1995; 3:739-45.
46. Lenschow DJ, Ho SC, Sattar H et al. Differential effects of anti-B7-1 and anti-B7-2 MAb treatment on the development of diabetes in the NOD mouse. J Exp Med 1995; 181:1145-55.
47. Kuchroo VK, Greer JM, Kaul D et al. A single TCR antagonist peptide inhibits experimental allergic encephalomyelitis mediated by a diverse T cell repertoire. J Immunol 1994; 153:3326-36.
48. Nicholson LB, Greer JM, Sobel RA et al. An altered peptide ligand mediates immune deviation and prevents experimental autoimmune encephalomyelitis. Immunity 1995; 3:397-405.
49. Karin N, Mitchell DJ, Brocke S et al. Reversal of experimental autoimmune encephalomyelitis by a soluble peptide variant of a myelin basic protein epitope: T cell receptor antagonism and reduction of interferon γ and tumor necrosis factor α production. J Exp Med 1994; 180:2227-37.
50. Pfeiffer C, Stein J, Southwood S et al. Altered peptide ligands can control CD4 T lymphocyte differentiation in vivo. J Exp Med 1995; 181:1569-74.
51. Lafaille JJ, Nagashima K, Katsuki M et al. High incidence of spontaneous autoimmune encephalomyelitis in immunodeficient anti-myelin basic protein T cell receptor transgenic mice. Cell 1994; 78:399-408.
52. Ferber IA, Brocke S, Taylor-Edwards C et al. Mice with a disrupted IFN-γ gene are susceptible to the induction of experimental autoimmune encephalomyelitis (EAE). J Immunol 1996; 156:5-7.
53. Ruddle NH, Bergman CM, McGrath KM et al. An antibody to lymphotoxin and tumor necrosis factor prevents transfer of experimental allergic encephalomyelitis. J Exp Med 1990; 1723:1193-200.
54. Chung IY, Norris JG, Benveniste EN. Differential tumor necrosis factor α expression by astrocytes from experimental allergic encephalomyelitis-susceptible and -resistant rat strains. J Exp Med 1991; 173:801-11.

55. Yednock TA, Cannon C, Fritz LC et al. Prevention of experimental autoimmune encephalomyelitis by antibodies against α4β1 integrin. Nature 1992; 356:63-6.
56. Kuchroo VK, Martin CA, Greer JM et al. Cytokines and adhesion molecules contribute to the ability of myelin proteolipid protein-specific T cell clones to mediate experimental allergic encephalomyelitis. J Immunol 1993; 151:4371-82.
57. Elfont RM, Griffin DE, Goldstein GW. Enhanced endothelial cell adhesion of human cerebrospinal fluid lymphocytes. Ann Neurol 1995; 38:405-13.
58. Nicholson LB, Waldner H-P, Carrizosa A et al. Superagonists: Altered peptides that hyperstimulate T cells and change their cytokine profile. Submitted.
59. Cannella B, Raine CS. The adhesion molecule and cytokine profile of multiple sclerosis lesions. Ann Neurol 1995; 37:424-35.
60. Khoury SJ, Hancock WW, Weiner HL. Oral tolerance to myelin basic protein and natural recovery from experimental autoimmune encephalomyelitis are associated with downregulation of inflammatory cytokines and differential upregulation of transforming growth factor β, interleukin 4, and prostaglandin E expression in the brain. J Exp Med 1992; 176:1355-64.
61. Kennedy MK, Torrance DS, Picha KS et al. Analysis of cytokine mRNA expression in the central nervous system of mice with experimental autoimmune encephalomyelitis reveals that IL-10 mRNA expression correlates with recovery. J Immunol 1992; 149:2496-505.
62. Renno T, Krakowski M, Piccirillo C et al. TNF-α expression by resident microglia and infiltrating leukocytes in the central nervous system of mice with EAE. J Immunol 1995; 154:944-53.
63. McRae BL, Kennedy MK, Tan LJ et al. Induction of active and adoptive relapsing experimental autoimmune encephalomyelitis (EAE) using an encephalitogenic epitope of proteolipid protein. J Neuroimmunol 1992; 38:229-40.
64. McRae BL, Vanderlugt CL, Dal Canto MC et al. Functional evidence for epitope spreading in the relapsing pathology of experimental autoimmune encephalomyelitis. J Exp Med 1995; 182:75-85.
65. Yu M, Johnson JM, Tuohy VK. A predictable sequential determinant spreading cascade invariably accompanies progression of experimental autoimmune encephalomyelitis: A basis for peptide-specific therapy after onset of clinical disease. J Exp Med 1996; 183:1777-88.
66. Nicholson LB, Murtaza A, Hafler BP et al. A T cell receptor antagonist peptide prevents autoimmune encephalomyelitis induced with multiple myelin antigens: Evidence for bystander suppression. Submitted.
67. Day MJ, Tse AGD, Puklavec M et al. Targeting autoantigen to B cells prevents the induction of a cell-mediated autoimmune disease in rats. J Exp Med 1993; 1992:655-9.

68. Saoudi A, Simmonds S, Huitinga I et al. Prevention of experimental allergic encephalomyelitis in rats by targeting autoantigen to B cells: Evidence that the protective mechanism depends on changes in the cytokine response and migratory properties of the autoantigen-specific T cells. J Exp Med 1995; 182:335-44.
69. Brocke S, Gijbels K, Allegretta M et al. Treatment of experimental encephalomyelitis with a peptide analogue of myelin basic protein. Nature 1996; 379:343-6.
70. Burstein HJ, Shea CM, Abbas AK. Aqueous antigens induce in vivo tolerance selectively in IL2 and IFN-γ producing (Th1) cells. J Immunol 1991; 148:3687-91.
71. Burstein HJ, Abbas AK. In vivo role of interleukin 4 in T cell tolerance induced by aqueous protein antigen. J Exp Med 1993; 177:457-63.
72. Khoruts A, Miller SD, Jenkins MK. Neuroantigen-specific Th2 cells are inefficient suppressors of experimental autoimmune encephalomyelitis induced by effector Th1 cells. J Immunol 1995; 155:5011-7.
73. Katz JD, Benoist C, Mathis D. T helper cell subsets in insulin-dependent diabetes. Science 1995; 268:1185-8.
74. Wogensen L, Lee M, Sarvetnick N. Production of interleukin-10 by islet cells accelerates immune-mediated destruction of cells in nonobese diabetic mice. J Exp Med 1994; 179:1379-84.
75. Lee M, Mueller R, Wicker LS et al. IL-10 is necessary and sufficient for autoimmune diabetes in conjunction with NOD MHC homozygosity. J Exp Med 1996; 183:2663-8.
76. Sarief MK, Hentges R. Association between tumor necrosis factor alpha and disease progression in patients with multiple sclerosis. N Engl J Med 1991; 325:467-72.
77. Tsukada N, Miyagi K, Matsuda M et al. Tumor necrosis factor and interleukin-1 in the CSF and sera of patients with multiple sclerosis. J Neurol Sci 1991; 102:230-4.
78. Maimone D, Gregory S, Arnason BGW et al. Cytokine levels in the cerebrospinal fluid and serum of patients of patients with multiple sclerosis. J Neuroimmunol 1991; 32:67-74.
79. Sharief MK, McLean B, Thompson EJ. Elevated serum levels of tumor necrosis factor in Guillain-Barre syndrome. Ann Neurol 1993; 33:591-6.
80. Astroianni CM, Paoletti F, Valenti C et al. Tumor necrosis factor and neurological disorders in HIV infection. J Neurol Neurosurg Psychiatry 1992; 55:219-21.
81. Franciotta DM, Grimaldi LME, Martino GV et al. TNF in serum and CSF of patients with multiple sclerosis. Ann Neurol 1989; 26:787-9.
82. Perrella O, Carrieri PB, De Mercato R et al. Markers of activated T-lymphocytes and T-cell receptor gamma delta+ in patients with MS. European Neurology 1993; 33:152-5.
83. Peter JB, Boctor FN, Tourtellotte WW. Serum and CSF levels of IL-2, sIL-2R, TNF-α and IL-1 beta in chronic progressive multiple sclerosis. Neurology 1991; 41:121-3.

84. Konttinen YT, Kinnunen E, Kemppinen P et al. Lymphocyte activation in discordant multiple sclerosis twin pairs. J Neuroimmunol 1990; 27:1-8.
85. Gallo P, Piccino MG, Krzalic L et al. Tumour necrosis factor and neurological diseases: Failure in detecting TNF in the CSF from patients with multiple sclerosis, AIDS dementia complex and brain tumor. J Neuroimmunol 1989; 23:41-4.
86. Mokhtarian F, Shi Y, Shirazian D et al. Defective production of anti-inflammatory cytokine TGF beta by T cell lines of patients with multiple sclerosis. J Immunol 1994; 152:6003-10.
87. Rieckmann P, Albrecht M, Kitze B et al. Cytokine mRNA levels in mononuclear blood cells from patients with multiple sclerosis. Neurology 1994; 44:1523-6.
88. Rieckmann P, Albrecht M, Kitze B et al. Tumor necrosis factor α messenger RNA expression in patients with relapsing remitting multiple sclerosis is associated with disease activity. Ann Neurol 1995; 37:82-8.
89. Zipp F, Weber F, Huber S et al. Genetic control of multiple sclerosis: Increased production of lymphotoxin and tumor necrosis factor-α by HLA-DR2$^+$ T cells. Ann Neurol 1995; 38:723-30.
90. Correale J, Gilmore W, McMillan M et al. Patterns of cytokine secretion by autoreactive proteolipid protein specific T cell clones during the course of multiple sclerosis. J Immunol 1995; 154:2959-68.
91. Brod SA, Benjamin DS, Hafler DA. Restricted T cell expression of IL-2/IFN-γ mRNA in human inflammatory disease. J Immunol 1991; 147:810-5.
92. Olsson T. Cytokines in neuroinflammatory disease: Role of myelin autoreactive T cell production of interferon-gamma. J Neuroimmunol 1993; 40:211-8.
93. Inoue T, Asano Y, Matsuka S et al. Distinction of mouse CD8$^+$ suppressor effector cell clones from cytotoxic T cell clones by cytokine production and CD45 isoforms. J Immunol 1993; 150:2121-8.
94. Antel JP, Bania MB, Noronha A et al. Defective suppressor cell function mediated by T8$^+$ cell lines from patients with progressive multiple sclerosis. J Immunol 1986; 137:3436-9.
95. Antel JP, Bania MB, Reder A et al. Activated suppressor cell dysfunction in progressive multiple sclerosis. J Immunol 1986; 137:137-41.
96. Chouffon M, Weiner HL, Morimoto C et al. Loss of functional suppression is linked to decreases in circulating suppressor inducer (CD4$^+$2H4$^+$) T cells in multiple sclerosis. Ann Neurol 1988; 24:185-91.
97. Hafler DA, Buchsbaum M, Weiner HL. Decreased autologous mixed lymphocyte reaction in multiple sclerosis. J Neuroimmunol 1985; 9:339-47.
98. Balashov KE, Khoury SJ, Hafler DA et al. Inhibition of T cell responses by activated human CD8$^+$ T cells is mediated by interferon-γ and is defective in chronic progressive multiple sclerosis. J Clin Invest 1996; 95:2711-9.

99. Traugott U, Lebon P. Interferon gamma and Ia antigen are present on astrocytes in active chronic multiple sclerosis lesions. J Neurol Sci 1988; 84:257-64.
100. Hofman FM, Hinton DR, Johnson K et al. Tumor necrosis factor identified in multiple sclerosis brain. J Exp Med 1989; 170:607-12.
101. Selmaj K, Raine CS, Cannella B et al. Identification of lymphotoxin and tumor necrosis factor in multiple sclerosis lesions. J Clin Invest 1991; 87:949-54.
102. Wesselingh SL, Power C, Glass J et al. Intracerebral cytokine messenger RNA expression in acquired immunodeficiency syndrome dementia. Ann Neurol 1993; 33:576-82.
103. Nagano I, Nakamura S, Yoshioka M et al. Expression of cytokines in brain lesions in subacute sclerosing panencephalitis. Neurology 1994; 44:710-5.
104. Windhagen A, Newcombe J, Dangond F et al. Expression of costimulatory molecules B7-1 (CD80), B7-2 (CD86), and interleukin 12 cytokine in multiple sclerosis lesions. J Exp Med 1995; 182:1985-96.
105. Woodroofe MN, Cuzner ML. Cytokine mRNA expression in inflammatory multiple sclerosis lesions: Detection by non-radioactive in situ hybridization. Cytokine 1993; 5:583-8.
106. Tyor WR, Glass JD, Griffin JW et al. Cytokine expression in the brain during the acquired immunodeficiency syndrome. Ann Neurol 1992; 31:349-60.
107. Smith DR, Balashov KE, Hafler DA et al. Immune deviation following pulse cyclophosphamide treatment of multiple sclerosis: Increased IL-4 production and associated eosiophilia. Submitted.
108. Rieckmann P, Weber F, Gunther A et al. Pentoxifylline, a phosphodiesterase inhibitor, induces immune deviation in patients with multiple sclerosis. J Neuroimmunol 1996; 64:193-200.
109. Rhodes J, Chen H, Hall SR et al. Therapeutic potentiation of the immune system by costimulatory Schiff-base-forming drugs. Nature 1995; 377:71-5.
110. Brod SA, Marshall GD, Henninger EM et al. Interferon-β1b treatment decreases tumor necrosis factor-α and increases interleukin-6 production in multiple sclerosis. Neurology 1996; 46:1633-8.
111. Jacobs L, Goodkin DE, Rudick RA et al. Advances in specific therapy for multiple sclerosis. Current Opinion in Neurology 1994; 7:250-4.
112. Panitch HS, Hirsch RL, Haley AA et al. Exacerbation of multiple sclerosis in patients treated with gamma-interferon. Lancet 1987; i:893
113. Panitch HS, Hirsch RL, Schindler J et al. Treatment of multiple sclerosis with gamma interferon: Exacerbations associated with activation of the immune system. Neurology 1987; 37:1097-102.

114. Weiner HL, Mackin GA, Matsui M et al. Double-blind pilot trial of oral tolerization with myelin antigens in multiple sclerosis. Science 1993; 259:1321-4.
115. Trentham DE, Dynesius-Trentham RA, Orav EJ et al. Effects of oral administration of type II collagen on rheumatoid arthritis. Science 1993; 261:1727-30.
116. Fukaura K, Kent SC, Pietrusewicz MJ et al. Induction of circulating myelin basic protein and proteolipid protein-specific transforming growth factor-β1-secreting T cells by oral administration of myelin in multiple sclerosis patients. J Clin Invest 1996; 98:70-7.

CHAPTER 7

Th2 Cytokines in the Treatment of Autoimmune Diseases: A Delicate Balance

Ari Waisman, Stefan Brocke and Lawrence Steinman

Cytokines are known to play an important role in autoimmune diseases. It is well established that in inflammatory autoimmune diseases of the central nervous system (CNS), such as multiple sclerosis (MS) in humans and experimental autoimmune encephalomyelitis (EAE) in animals, cytokines of the Th1 type play an important role. Th1 cytokines, including IL-12, IFN-γ and the pro-inflammatory cytokine TNF-α were found in the CNS of MS patients and of animals with EAE. Systemic administration of cytokines has paradoxical effects at times in the EAE model. For example, although IFN-γ is associated with worsening of multiple sclerosis, administration of IFN-γ improves EAE. In contrast, utilization of methods that change the overall balance of cytokines, from Th1 to Th2 are much more predictable and reliable in modifying disease. In this review we will discuss novel techniques aimed at protection from autoimmune diseases. These techniques, which include the use of altered peptide ligands, and DNA vaccination against the T cell receptor, shift the overall balance between Th1 and Th2 cytokines, and protect experimental animals from autoimmune disease, and at times even reverse ongoing clinical deficits.

INTRODUCTION

It is well established that cytokines play an important role in the immune response. T cells were found to secrete many different cytokines, and to be influenced by many others. Clones of $CD4^+$ or $CD8^+$ T cells are subtyped based on their spectrum of cytokine expression, with a striking correlation to their functional capacities.[1] T cells that secrete primary IFN-γ and IL-12 are termed Th1, while T cells that secrete mostly IL-4, IL-5, IL-10 and IL-13 are named

Immunointervention in Autoimmunity by Th1/Th2 Regulation, edited by Luciano Adorini.

Th2. Th1 cells function as T effector cells involved in delayed type hypersensitivity (DTH) and also supply help to B cells for the Ig class shift to IgG2a, while Th2 cells mainly provide help to B cell proliferation and differentiation. While T cell clones that are cultured long-term in vitro can usually be divided to Th1, Th2 or Th0 (cells that produce cytokines of both Th1 and Th2) sub-types, this division is not always so clear when short-term cultures or T cell lines are analyzed.

Multiple sclerosis (MS) is a chronic inflammatory disease of the central nervous system (CNS). The disease is characterized by a demyelination in the white matter of the brain and spinal cord. Lesions in the myelin sheath contain T cells, B cells, macrophages and immunoglobulin. Pro-inflammatory cytokines, such as IFN-γ and TNF-α are suspected to be involved in disease development, and will be discussed in length. Experimental autoimmune encephalomyelitis (EAE) is a disease that can be induced in rodents and nonhuman primates, and serves as a model for MS. EAE can be induced either by active immunization of the animal with myelin constituents, the most effective being MBP, proteolipid protein (PLP) and myelin oligodedndrocyte glycoprotein (MOG). The immunogen is mixed with complete Freund's adjuvant (CFA) and is injected into the animal, which develops clinical paralysis, typically after 10-14 days. T cell lines and clones that are specific to myelin antigens can also be used to induce the disease. The clones or lines are raised in culture, and when found to be specific for the target peptide or protein, injected to recipient animals. When such T cell lines and clones are tested for their cytokine secretion pattern, they are usually found to secrete Th1 cytokines, IFN-γ and IL-2, together with the pro-inflammatory cytokine TNF-α.

EAE was found to be a Th1 disease by other criteria as well. For example, following the immunization of SJL mice with the immunodominant peptide MBP87-99 in CFA, the lymph node cells of the mice, tested 10 days later, are found to secrete primarily Th1 cytokines in response to the immunogen. When such cells are then grown for longer periods in culture, they normally secrete Th1 cytokines, and are capable of inducing EAE in naive animals by passive transfer.

CYTOKINES IN MS AND EAE

Since MS is a neurological disease, cytokines may play a role in three major ways (see Table 7.1). First, they could be involved in the triggering of pathogenic T cells. Second, they could direct the T cells to the brain, and third, they could activate the T cells at the site of inflammation. These cytokines may be produced either by the T cells, or by neighboring cells such as antigen presenting cells, including astrocytes, microglial cells, and perhaps oligodendroglial cells, as well as B cells.

There is considerable evidence that the pathogenesis of MS is associated with Th1 and pro-inflammatory cytokines. On the other hand, there are many reports demonstrating that Th2 cytokines, primarily IL-4 and IL-10, and TGF-β (that may be associated with another group of T helper cells, Th3), are involved in the process of down-regulating disease.

The major cytokines involved in the pathogenesis of MS and EAE, are TNF-α and IFN-γ.[2] When cytokine mRNA levels were measured in mononuclear blood cells from MS patients, a significant association was found between the levels of TNF-α and disease activity.[3,4] Using immunocytochemical techniques, the presence of TNF was analyzed in the brain of MS patients and controls. In brain lesions in MS, and subacute sclerosing panencephalitis, TNF^+ cells were demonstrated, in association with both astrocytes and macrophages. These observations were not seen in Alzheimer's disease brain or normal brain tissue.[5] In addition, IFN-γ has been localized to MS lesions with immunohistochemistry.[6] Moreover, there are increased numbers of cells secreting IFN-γ among CSF of MS patients in comparison to controls.[7] When such cells were analyzed for their IFN-γ mRNA levels, such an elevation was found.[8] IFN-γ was tried as a treatment for MS, and induced clinical exacerbations; this led to a cessation of the trial and indicated that this an important cytokine in the disease (see Table 7.1).[9]

When sera and CSF of active MS patients was analyzed, increased levels of TNF-α were detected, in comparison to normal or noninflammatory neurological disease controls.[10,20,21] IL-1β, TNF-α and IL-6 were measured by specific radioimmunoassays in the CSF of patients with MS and other neurologic diseases. There was a high incidence of detectable IL-1β and TNF-α, but not IL-6 in patients with active MS compared with inactive MS or other neurologic diseases patients.[10] As IL-1β and TNF experimentally induce astrogliosis, demyelination, temperature elevation, lassitude, and sleep, the authors raise the possibility that these cytokines may contribute to a variety of manifestations in MS and in other disease states.[10]

Cytokine analysis was done in the rat at various stages in the development of EAE, using their spinal cord as the RNA source. It was found that cytokines associated with Th1 cells were elevated during the phase of the disease where the rats were paralyzed or died.[18] Similarly, when Olsson and collaborators analyzed the mRNA pattern in the CNS of Lewis rats, they found that the Th1 cytokines IL-12 and TNF-β appeared before clinical signs of disease, favoring a role for those cytokines in disease initiation, which was followed by the elevation in the levels of TNF-α.[19] Only when disease remitted, were high levels of IL-10 noted.[19] When the same group analyzed the expression of other cytokines by in situ methods, they found that, similar to their previous results, IFN-γ peaked before disease was apparent, and IL-4 mRNA appeared only during

Table 7.1. Cytokines involved in the pathogenesis of MS and EAE

Cytokine	Function	Findings	Reference
IL-1β	Pro-inflammatory	Elevated levels in CSF of MS patients	10
IL-2	T cell growth factor	High levels in serum of active	11, 12
		mRNA upregulated in the CNS of mice with EAE	13
IL-4	Th2	mRNA upregulates during remission of MS	14
IL-5	Th2	High levels in T cells isolated from MS patients treated by oral tolerance	15
IL-6	Pro-inflammatory	Elevated in CNS of mice with EAE	16
IL-10	Th2; Suppressive	Elevated in mononuclear cells of MS patients	17
		High level correlates with recovery from EAE	18
		Elevated when EAE remits in rats	19
IL-12	Th1	Elevated in CNS of rats with EAE	19
IL-13			
IFN-γ	Th1; Pro-inflammatory	Immunochemistry shows presence in brain of MS	6
		Elevated in CNS of EAE	16
		High number of secreting cells in CSF	7
TNF-α	Pro-inflammatory	Association between level of mRNA in blood and disease	3, 4
		mRNA transcripts in MS brain	5
		Increased levels in sera	20, 21
		Elevated levels in CNS of EAE	16
		Identification in MS lesions	22
		Elevated levels in CSF of MS patients	10, 23
TNF-β	Suppressive cytokine	Elevated in CNS of rats before EAE appears	19
sIL-2R	Cytokine receptor	Elevated in serum and CSF of MS patients	12, 24
sTNF-α	Cytokine receptor	Increased levels in sera	25

remission.[14] Taken together, those results demonstrate that in MS, Th1 cytokines participate in disease induction, which is followed by appearance of pro-inflammatory cytokines. Only at remission are Th2 cytokines such as IL-4 and IL-10, together with the suppressive TGF-β, present.

In mice, levels of IL-2 and IFN-γ mRNA in CSF cells were significantly elevated in mild EAE and strongly upregulated in severe disease, concordant with levels of those cytokines in total CNS tissue. These results confirm the CSF as representative of the immune status of the CNS, and indicate a role for IL-2 and IFN-γ in inflammatory CNS disease.[13,26]

Another way to analyze the cytokine pattern in MS is to analyze lymphocytes isolated from patients and animals. T cells directed to MBP peptides, isolated from MS patients, were normally found to be cytotoxic.[27] The latter may explain some of the mechanism by which those $CD4^+$ T cells may induce damage in the brain. When tested for cytokine secretion, those cells were found to secrete many Th1 and pro-inflammatory cytokines, although not exclusively.[28,29] Moreover, when T cell clones specific to PLP, isolated from MS patients with active disease were analyzed, it was found that most of them were Th1, namely secreting IFN-γ and TNF, but not IL-4, IL-10 or TGF-β.[29] Notably, when similar clones were isolated from patients in remission, many of those cells were able to produce IL-10 and TGF-β,[29] showing again the importance of those two cytokines in disease remission.

Similar studies were first conducted in rats by Ben-Nun, Wekerle and Cohen,[30] who demonstrated for the first time that T cells alone are capable of inducing EAE.[31] When pathogenic T cell lines and clones were analyzed for cytokine secretion, they were found to secrete manly Th1 cytokines.[32,33] The most important cytokine was found to be TNF-α. When a panel of T cell clones directed to the peptide MBPAc1-11 were analyzed, a strong correlation was found between pathogenicity and TNF-α production.[34] Namely, the clones that induced the strongest disease also produced the highest amounts of TNF-α, while clones that did not induce disease did not produce TNF-α.

TREATMENT OF MS AND EAE WITH CYTOKINES

At least three different cytokines are strong candidates for downregulating autoimmune inflammation in the central nervous system, perhaps partly by counteracting effects and production of IFN-γ. These anti-inflammatory cytokines are TGF-β, IL-4 and IL-10 (see Table 7.2). As mentioned above, IL-10 and TGF-β transcripts are found in the brain of MS patients and EAE animals in remission.

Of the Th2 cytokines, the most intensively investigated is IL-4. However, its role in MS and EAE is not clear. No direct correlation was found between clinical recovery and its levels in rats with EAE.[14,18,19] IL-4 is produced by Th2 cells, and is also used by the immune system to suppress Th1 responses. IL-4 induces T helper cells secreting Th1 and Th2 cytokines (Th0), to produce only or mainly Th2 cytokines. Since EAE is a Th1 disease, it was thought that systemic introduction of this important cytokine may have a beneficial outcome. Indeed, it was found that mice that systematically received IL-4 developed a milder form of EAE, as compared to control animals.[35] In addition, long-term injection with IL-4, prior to EAE induction, abolished the disease.[36] IL-4 has, however, a paradoxical role in EAE that can only be compared with that of IFN-γ, discussed later. There is evidence that this cytokine can also worsen EAE, depending on the timing of treatment.[16] We hypothesize that since IL-4 receptor is present also on Th1 cells, its administration may activate the latter, worsening disease (see Table 7.2).

To assess the effects of IL-10 on different rat lymphoid cell functions in vitro and to consider the possibility of an IL-10-mediated treatment to prevent the induction of CNS autoimmune disease in vivo, it was systemically administered during the initiation phase of disease, and was effective in markedly suppressing the subsequent induction of EAE in Lewis rats.[38] This suppression of clinical disease coincided with a significant and specific elevation of MBP-specific autoantibody production, a sustained T cell proliferative response to MBP and a diminution of CNS infiltration and thymic involution in diseased animals.[38]

IL-13 is a pleiotropic cytokine that shares many of the characteristics of IL-4. IL-13 is produced by Th2 cells.[53] In addition, like IL-4, IL-13 can inhibit the expression of pro-inflammatory cytokines such as IL-1β, TNF-α and IL-6.[53] The latter may be the basis for its strong influence in EAE. It was found that upon in vivo application of hrIL-13-secreting cells into MBP-immunized animals, the cytokine was capable of markedly suppressing the development of EAE, as assessed by a reduction of the mean duration, severity, and incidence of disease.[39]

The role of TGF-β in the suppression of EAE and MS is well defined. TGF-β is the main cytokine found to be involved in the process of oral tolerance of EAE (see separate section). As described previously, its presence correlates with remission in MS patients, and with amelioration of EAE. A direct role for that cytokine was found in suppression of secretion of Th1 and pro-inflammatory cytokines by T cells specific to MBP and to PLP isolated from MS patients.[54] In contrast, the secretion of IL-10 and TGF-β itself by those cells was not effected.[54]

TGF-β was injected at different intervals after the EAE-inducing immunization. Interestingly, it was found that TGF-β is protective only if given in the last 5 days before disease appears, but not if given earlier or after disease already had started.[51] In agreement, it was found that anti-TGF-β accelerates and aggravates EAE when given just before disease appears, but not if given after.[51] Clinical trials of TGF-β are presently being undertaken in the laboratory of Dr. Henry McFarland at NIH.

Table 7.2. Cytokines used in treatment of EAE

Cytokine	Route	Findings	Reference
IL-1α	i.p. injection	Inhibition of EAE	37
IL-1β	In virus construct	Inhibition of EAE	16
IL-2	In virus construct	Inhibition of EAE	16
IL-4	i.v. injection In virus construct	Amelioration of disease Worsening of EAE	35 16
IL-6	In virus construct	Inhibition of EAE	16
IL-10	In virus construct Systemic administration	Inhibition of EAE Prevention of EAE	16 38
IL-13	Administration of vector cells	Suppression of disease	39
IFN-α	Systemic administration	No significant effect in MS	40, 41
IFN-α2a	Systemic administration	Reduces MS clinical signs	42
IFN-β	Systemic administration	Significant beneficial effect on MS	43-47
IFN-γ	Systemic administration Local administration	Worsening of MS Amelioration of EAE in rat	9 48
TNF-α	Systemic administration	Augments EAE in rats	49
TGF-β	Systemic administration	Suppression of disease	50-52

Since IFN-γ is an important Th1 cytokine, and is also found in lesions in CNS of MS patients and animals with EAE (see above), it was hypothesized that antibodies that neutralize it would have a beneficial role in EAE. To that end, anti-IFN-γ antibodies were used to treat SJL mice in which EAE was induced either passively or actively. Paradoxically, these antibodies induced a stronger form of the disease.[55] In agreement, treatment of the C57BL/6 mice, which are low responders to EAE, by anti-IFN-γ antibodies resulted in an increased percentage of paralyzed mice.[56]

The therapeutic potential of IFN-γ in human MS was tested by treatment of clinically definite relapsing-remitting MS patients with recombinant IFN-γ. It was found that 7 of the 18 patients had an exacerbation during treatment, a significant increase compared with the prestudy exacerbation rate. These results showed that systemic administration of IFN-γ has pronounced effects on cellular immunity in MS and on disease activity within the CNS, suggesting that the attacks induced during treatment were immunologically mediated.[9]

In sharp contrast to the pathologic effect of IFN-γ in MS, the effect of IFN-γ on EAE in Lewis rats was beneficial. Systematically administered IFN-γ did not change the disease course of EAE, whereas IFN-γ applied locally into the ventricular system of the CNS resulted in complete suppression of clinical signs.[48] Furthermore, it was found that systemic administration of anti-IFN-γ just prior to the onset of clinical symptoms resulted in a more severe disease course.[48]

IL-12 is a novel heterodimeric cytokine that enhances NK-mediated cytotoxicity and induces IFN-γ production by NK cells and T cells.[57-59] IL-12 is an important Th1 cytokine since its capable of promoting Th1-specific immune responses.[60-62] When EAE was induced with PLP-specific LNC that were stimulated in the presence of IL-12, the course of the disease was more severe and prolonged.[63] The latter may have arisen from a two-fold increase in the level of TNF-α produced by those cells.[63] In contrast, treatment of the mice with antibodies to IL-12 prior to LNC transfer resulted in a milder form of EAE,[63] demonstrating that endogenous IL-12 plays a pivotal role in the pathogenesis of the disease.

TREATMENT OF MS AND EAE WITH CYTOKINE INHIBITORS

Cytokines such as interleukin-1 (IL-1) are thought to contribute to the inflammatory response associated with MS and EAE. Treatment with an IL-1 receptor antagonist (recombinant human IL-1ra) in Lewis rats, following EAE induction with MBP, resulted in delayed disease onset, reduction in the severity of paralysis and weight loss, with shortening of the duration of disease.[64] These data suggest that IL-1 is a mediator of the inflammation resulting from active immunization with MBP, and that inhibitors of IL-1 may prove beneficial for the treatment of autoimmune or inflammatory diseases of the central nervous system.

TNF-α is an important pro-inflammatory cytokine in MS and in EAE. Suppression of TNF-α for treatment of EAE has been tried in several different formats. One such inhibitor is a dimeric polyethylene glycol linked form of the type I soluble receptor of TNF, PEG-$(rsTNF\text{-}RI)_2$. When PEG-$(rsTNF\text{-}RI)_2$ was used to treat rats that were immunized with MBP, EAE clinical signs were significantly inhibited in a dose-dependent manner.[65] Histological examination of the central nervous system indicated that the administration of PEG-(rsTNF-RI)$_2$ reduced, in part, the cellular infiltrate, particularly in the lumbar and sacral regions of the spinal cord.[65] These studies suggest that TNF is a pivotal mediator of the inflammation resulting from the complete immune response induced by active immunization with MBP (see Table 7.3).

Rolipram, an anti-depressant used in the clinic, was tested for its efficiency in the treatment of EAE. It was found that rolipram, when incubated with Th1 cells isolated from MS patients, lowered the level of TNF-α secreted by those cells [66]. It was therefore tested in for its capability to inhibit EAE, and was found to inhibit the disease, if injected daily.[66]

TNF-α is an important mediator in the pathogenesis of MS and EAE. Further support for the central role of TNF-α in the disease pathogenesis was shown by the use of monoclonal antibodies directed to TNF-α, in an attempt to neutralize the activity of the cytokine in the CNS. It was found that neutralization of TNF-α activity within the CNS exhibits significant improvement in disease[69,70] Furthermore, local administration of anti-TNF-α monoclonal antibodies had a greater effect on ameliorating disease, in comparison to systemic administration of the antibodies,[69] those reinforcing the importance of inhibition of cytokines in the site of inflammation. Methods that inhibit TNF-α production at the site of disease could be used in the future for the therapy of MS.[72]

Advanced clinical trials have been performed using IFN-β. This cytokine was found, similarly to TGF-β and IL-10, to be involved in the suppression of

Table 7.3. Cytokine inhibitors used in treatment of EAE

Cytokine	Inhibitor	Findings	Reference
IL-2	Anti-IL-2	Decrease EAE	55
IL-12	Anti-IL-12	Prevent EAE	63
IFN-γ	Anti-IFN-γ	Worsening of EAE	48, 55, 56
TNF-α	TNF-αR-IgG	Prevent EAE signs	67
	Anti-TNF-α	Prevents I-CAM and V-CAM expression, reduces EAE	68
	Anti-TNF-α	Prevent and suppress EAE	69
	Anti-TNF-α	Prevent transfer of EAE	70
	TNFα receptor	Prevent and suppress EAE	69
sIL-1R	IL-1 receptor	Reduces EAE duration and severity	37
MIP1α	Anti-MIP1α	Treats and prevent EAE	71

EAE. In addition, it was found that in vitro IFN-β can reduce the production of IFN-γ.[73] When tested in relapsing-remitting MS patients, it was found that systemic administration of IFN-β could reduce the number of relapses in MS, and decrease lesion activity on magnetic resonance scans.[45-47] These results have lead to the licensing of this drug for clinical use in the USA for patients with relapsing remitting MS. The most likely mechanism of action of IFN-β may be via reduction of the expression of IFN-γ in the CNS.

ORAL TOLERANCE OF AUTOIMMUNE DISEASES

Oral administration of antigen is used to induce antigen-specific peripheral immune tolerance, a process of immunologic hyporesponsiveness. As well as preventing systemic immune responses to ingested proteins, oral tolerance to autoantigens has also been used to suppress autoimmune diseases in animals and humans.[74,75] It was shown that different mechanisms are involved in mediating oral tolerance depending on the dose fed. Low doses of antigen generate cytokine-secreting regulatory cells, whereas high doses induce anergy or deletion.[74,75]

Oral antigen was found to delete antigen-reactive T cells in Peyer's patches, in mice transgenic for the ovalbumin-specific T cell receptor genes. The deletion was mediated by apoptosis, and was dependent on dosage and frequency of feeding. At lower doses deletion was not observed; instead there was induction of antigen-specific cells that produced TGF-β and IL-4 and IL-10 cytokines. At higher doses, both Th1 and Th2 cells were deleted following their initial activation, whereas cells which secrete TGF-β were resistant to deletion.[76]

Using mice transgenic for a TCR derived from an encephalitogenic T cell clone specific for the peptide MBPAc1-11 plus I-A^u, the question of whether regulatory T cells could be generated from the same precursor cell as that of an encephalitogenic Th1 cell was tested. The MBP TCR transgenic mice primarily have T cells of a precursor phenotype, Th0. Upon feeding with mouse MBP it was found that cells producing Th2 cytokines and TGF-β were formed, but only if low levels of antigen were used for the feeding. In contrast, high level feeding did not alter the cytokine pattern.[77]

T cell clones were isolated from the mesenteric lymph nodes of SJL mice that had been orally tolerized to MBP were CD4$^+$ and were structurally identical to Th1 encephalitogenic CD4$^+$ clones in T cell receptor usage.[33] In contrast to the latter, they produced TGF-β, IL-4 and IL-10 and suppressed EAE induced with either MBP or PLP. Thus, mucosally derived Th2-like clones induced by oral antigen can actively regulate immune responses in vivo and may represent a different subset of T cells.[33]

THE ROLE OF COSTIMULATORY MOLECULES

Costimulatory molecules are known to have an important role in the activation of CD4 T helper cells. Two costimulatory molecules on the surface of the APC, B7-1 and B7-2, were found to interact with the CD28 molecule on the T cell. It was found that these two different molecules transmit differential signals to T cells. In plaques taken from MS brains, it was found that B7-1, coexpressed with the mRNA for IL-12, was more frequent then the expression of B7-1.[78] In concordance, it was shown that when T cells are activated with the aid of B7-1 it produces Th1 cytokines, while Th2 cytokines are produced following interaction with B7-2.[79] Together, those two papers demonstrate that those B7-1 and B7-2 can transmit different signals that lead to Th1 and Th2 cytokine patterns, respectively. Using anti-B7 antibodies, this developmental step was manipulated both in vitro and in vivo in EAE. Anti-B7-1 reduced the incidence of disease while anti-B7-2 increased disease severity, in both cases not effecting the efficiency of the response to the antigen, but only the cytokine pattern.[79]

Similar work was done on the contribution of the CD28-B7 costimulatory T cell activation pathway to the pathogenesis of EAE in the Lewis rat model. Systemic administration of CTLA4Ig, which blocks the CD28-B7 pathway suppressed clinical disease, and was effective even when CTLA4Ig was delayed until day 10 postimmunization, a time when pathologic disease is evident.[80] Detailed immunohistologic studies showed that CTLA4Ig therapy resulted in suppression of the inflammatory response with inhibition of Th1 (IL-2 and IFN-γ) and sparing of Th2 (IL-4, IL-10, and IL-13) cytokines in the central nervous system.[80] As Khory and colleagues explain, "these results indicate that the CD28-B7 T cell costimulatory pathway plays an important role in EAE, and suggests that blockade of this costimulatory pathway protects against active disease by causing a state of immune deviation towards Th2 function. The ability of CTLA4Ig to treat animals with pathologically established disease may have important clinical implications for patients with MS".[80]

DNA VACCINATION

The search for specific immunotherapy for autoimmune diseases mediated by pathogenic T cells has focused on the development of strategies which target the T cell receptor.[81] In 1981 Ben-Nun, Wekerle and Cohen established the utility of vaccinating animals with pathogenic T cells capable of causing EAE in order to elicit anti-T cell immunity and treat clinical disease.[82,83] Vandenbark and colleagues and Brostoff and colleagues showed that immunization with peptides spanning either the CDR2 or CDR3 regions of the T cell receptor beta chain, which recognizes the dominant epitope of myelin basic protein, would protect animals from EAE. T cell vaccination[84,85] and immunization with CDR peptides[15,86] have been attempted in humans with multiple sclerosis, in early stage clinical trials with some indications of success.[84]

We have previously shown that the response to MBPAc1-11 dominates the immune response to MBP. Thus when Vβ8.2 bearing T cells are deleted with a monoclonal antibody targeting the Vβ8.2 gene product, EAE can be prevented and even reversed when disease is induced by the entire MBP molecule.[87]

Immunization with DNA has proven successful in obtaining immunity to microbes and to tumor antigens.[88-91] Since DNA vaccination has became so popular, a special site was formed for it on the world wide web: http://www.genweb.com/Dnavax/dnavax.html.

We have demonstrated the feasibility of DNA vaccination to a T cell receptor V gene, Vβ8.2, that is commonly found on pathogenic T cells which cause EAE in H-2^u mice. This approach prevents disease when it is elicited with MBPAc1-20.[92] T cells with specificity for MBPAc1-20 predominately rearrange and express the Vβ8.2 product in their TCR. This approach also prevented disease, though to a lesser degree, when EAE was induced with the entire MBP molecule. The response to MBP is more diverse, involving immunity to several epitopes where Vβ8.2 is not involved in T cell recognition.

We have demonstrated that DNA vaccination with a TCR V gene induces a shift in the pattern of cytokine production in T cells that utilize the relevant V gene segment. Thus, after TCR V gene vaccination the immune response to the dominant epitope on MBP shifts from Th1 to Th2. This observation is corroborated by a shift in vivo in the isotype of the antibodies produced against the immunodominant antigen MBPAc1-20.[92] The development of Th2 responses toward MBPAc1-20 may allow the prevention of EAE without the need for depletion of Vβ8.2 bearing T cells. Moreover, these Th2 cytokines may prevent EAE when whole MBP is used as the immunogen, and determinant spreading to other epitopes than MBPAc1-20 has occurred.[93]

A shift from Th1 to Th2 immunity has not been seen with DNA immunization towards microbial antigens. For example, immunization with pCMV/gp63 DNA, encoding the *Leishmania major* surface glycoprotein, produced T cells making significant amounts of IL-2, γ-interferon, but not IL-4 in response to *L. major*.[94] Clearly, immunization with DNA encoding TCR V genes induces alterations in the immune response that are not seen following DNA immunization with other antigens. Whether the shift from Th1 to Th2 immunity is uniquely seen with DNA vaccination to TCR genes is under investigation. Preliminary experiments indicate that vaccination with TCR Vβ8.2 changes the cytokine profile of T cells recognizing sperm whale myoglobin peptide 110-121, recognized predominately by Vβ8.2 T cells, from a Th1 to a Th2 response.[92]

Thus, DNA vaccination with TCR constructs may induce changes in T cells responding to foreign antigens, and not merely in T cells responding to autoantigens like MBP. In contrast the cytokines produced by T cells recognizing PPD are not changed.[92] PPD is not recognized by T cells predominately expressing Vβ8.2.[87] Thus, the effect may be restricted to T cells utilizing the TCR Vβ gene used in the vaccination.

DNA vaccination to a dominant T cell receptor V gene found on T cells recognizing a critical pathogenic determinant on a myelin protein may be useful in MS. Certain TCR V genes are expanded clonally in the T cell response to the immunodominant portion of the MBP molecule.[84,85,95-99] It would be possible to target the TCR V genes expressed in these clonally expanded T cells with DNA vaccination. This strategy has the added advantage that the response to the target antigen may then modulate from Th1 to Th2 and thus provide a suppressive environment for the responses to other myelin components that occur in the MS plaque.[97,100,101]

TREATMENT OF AUTOIMMUNE DISEASES BY ALTERED PEPTIDE LIGANDS

Characterization of the T cell responses to altered peptide ligands (APL) have provided functional evidence that a T cell receptor can distinguish subtle structural changes in its ligand, stressing the complexity of this antigen receptor signaling system. Over the past year, observations from many studies have suggested several roles for such analog peptides in various aspects of immune responses.

In autoimmunity, various studies were done, mostly utilizing T cells that induce EAE, or T cells isolated from MS patients. In EAE induced with myelin proteolipid protein (PLP) peptide 139-151, an APL generated by a single amino acid substitution (tryptophan to glutamine) at position 144 (Q144), inhibits the development of EAE induced with the native PLP 139-151 peptide (W144).[102] Adoptive transfer of T cell lines generated with the APL confer protection from EAE.[102]

The peptide p85-99 of MBP was used to alter the cytokine secretion pattern of human TCL. Using the APL, it was demonstrated that a Th0 T cell clone no longer produced IL-2, IL-4, IL-10, and IFN-γ, but instead secreted the suppressive cytokine TGF-β1.[103]

APL built around the same region of MBP, between residues 85-100, were used to clarify the role of APL in both rat and mice. Although not the dominant peptide of MBP in Lewis rat,[104] the peptide p87-99 can nevertheless induce significant EAE when injected in CFA.[105] It was found that when the lysine in position 91 of the peptide p87-99 was replaced with an alanine, the altered peptide could not induce active EAE, although it could trigger the proliferation of a T cell line specific to the native p87-99.[106] It was found that when administrated in a soluble form this APL could inhibit EAE induced by a pathogenic TCL specific to the native peptide. APL treatment was followed by a reduction in the secretion of the pro-inflammatory cytokines IFN-γ and TNF-α by lymph node cells isolated from the treated animals.[106] Those results demonstrate that when given in vivo, an altered peptide ligand of a pathogenic peptide can inhibit autoimmune disease.

In SJL mice, the MBP peptide p87-99 is dominant, and upon immunization to naive mice, induces EAE.[105] Following induction of experimental encephalomyelitis with a T cell clone, L10C1, that is specific for the myelin basic protein epitope p87-99, the inflammatory infiltrate in the central nervous system contains a diverse collection of T cells with heterogeneous receptors. We have shown that when clone L10C1 is tolerized in vivo with an analogue of p87-99, established paralysis is reversed, inflammatory infiltrates regress, and the heterogeneous T cell infiltrate disappears from the brain, with only the T cell clones that incited disease remaining in the original lesions.[95] We found that antibody raised against IL-4 reversed the tolerance induced by the altered peptide ligand.[95] Treatment with this altered peptide ligand selectively silences pathogenic T cells and actively signals for the efflux of other T cells recruited to the site of disease as a result of the production of IL-4 and the reduction of TNF-α in the lesion.[95]

CONCLUSIONS

There is a large diversity in the immune response in MS. This realization makes it unlikely that highly specific treatment of the disease, targeting a single epitope or TCR V gene product, for example, would succeed. Nevertheless, it is possible to target a critical epitope or TCR V gene product in such a manner that the cytokine profile shifts from Th1 to Th2. When this occurs the highly specific therapy becomes "trans-acting" leading to the suppression of other immune responses occurring at the site of disease. This allows for a degree of specificity in the therapy, while providing for local changes in cytokines at the site of disease. Moreover, the potential toxicity of administering cytokines or their antagonists systemically is overcome. It is clear that manipulation of cytokines at the site of disease may well prove beneficial in the treatment of organ specific autoimmune diseases such as MS.

References

1. Mosmann TR, Coffman RL. TH1 and TH2 cells: different patterns of lymphokine secretion lead to different functional properties. Annu Rev Immunol 1989; 7:145-173.

2. Brosnan CF, Cannella B, Battistini L et al. Cytokine localization in multiple sclerosis lesions: correlation with adhesion molecule expression and reactive nitrogen species. Neurology 1995; 45:S16-21.
3. Rieckmann P, Albrecht M, Kitze B et al. Tumor necrosis factor-alpha messenger RNA expression in patients with relapsing-remitting multiple sclerosis is associated with disease activity. Ann Neurol 1995; 37:82-88.
4. Rieckmann P, Albrecht M, Kitze B et al. Cytokine mRNA levels in mononuclear blood cells from patients with multiple sclerosis. Neurology 1994; 44:1523-1526.
5. Hofman FM, Hinton DR, Johnson K et al. Tumor necrosis factor identified in multiple sclerosis brain. J Exp Med 1989; 170:607-612.
6. Traugott U, Lebon P. Multiple sclerosis: involvement of interferons in lesion pathogenesis. Ann Neurol 1988; 24:243-251.
7. Olsson T, Zhi WW, Hojeberg B et al. Autoreactive T lymphocytes in multiple sclerosis determined by antigen-induced secretion of interferon-gamma. J Clin Invest 1990; 86:981-985.
8. Link J, Soderstrom M, Olsson T et al. Increased transforming growth factor-beta, interleukin-4, and interferon-gamma in multiple sclerosis. Ann Neurol 1994; 36:379-386.
9. Panitch HS, Hirsch RL, Schindler J et al. Treatment of multiple sclerosis with gamma interferon: exacerbations associated with activation of the immune system. Neurology 1987; 37:1097-1102.
10. Hauser SL, Doolittle TH, Lincoln R et al. Cytokine accumulations in CSF of multiple sclerosis patients: frequent detection of interleukin-1 and tumor necrosis factor but not interleukin-6. Neurology 1990; 40:1735-1739.
11. Trotter JL, Collins KG, van der Veen RC. Serum cytokine levels in chronic progressive multiple sclerosis: interleukin-2 levels parallel tumor necrosis factor-alpha levels. J Neuroimmunol 1991; 33:29-36.
12. Sharief MK, Hentges R, Ciardi M et al. In vivo relationship of interleukin-2 and soluble IL-2 receptor to blood-brain barrier impairment in patients with active multiple sclerosis. J Neurol 1993; 240:46-50.
13. Renno T, Lin JY, Piccirillo C et al. Cytokine production by cells in cerebrospinal fluid during experimental allergic encephalomyelitis in SJL/J mice. J Neuroimmunol 1994; 49:1-7.
14. Issazadeh S, Mustafa M, Ljungdahl A et al. Interferon gamma, interleukin 4 and transforming growth factor beta in experimental autoimmune encephalomyelitis in Lewis rats: dynamics of cellular mRNA expression in the central nervous system and lymphoid cells. J Neurosci Res 1995; 40:579-590.
15. Chou YK, Morrison WJ, Weinberg AD et al. Immunity to TCR peptides in multiple sclerosis. II. T cell recognition of V beta 5.2 and V beta 6.1 CDR2 peptides. J Immunol 1994; 152:2520-2529.
16. Willenborg DO, Fordham SA, Cowden WB et al. Cytokines and murine autoimmune encephalomyelitis: inhibition or enhancement of disease

with antibodies to select cytokines, or by delivery of exogenous cytokines using a recombinant vaccinia virus system. Scand J Immunol 1995; 41:31-41.
17. Navikas V, Link J, Palasik W et al. Increased mRNA expression of IL-10 in mononuclear cells in multiple sclerosis and optic neuritis. Scand J Immunol 1995; 41:171-178.
18. Kennedy MK, Torrance DS, Picha KS et al. Analysis of cytokine mRNA expression in the central nervous system of mice with experimental autoimmune encephalomyelitis reveals that IL-10 mRNA expression correlates with recovery. J Immunol 1992; 149:2496-2505.
19. Issazadeh S, Ljungdahl A, Hojeberg B et al. Cytokine production in the central nervous system of Lewis rats with experimental autoimmune encephalomyelitis: dynamics of mRNA expression for interleukin-10, interleukin-12, cytolysin, tumor necrosis factor alpha and tumor necrosis factor beta. J Neuroimmunol 1995; 61:205-212.
20. Carrieri PB, Maiorino A, Provitera V et al. Cytokines in the pathogenesis of multiple sclerosis. Acta Neurol (Napoli) 1992; 14:333-341.
21. Sharief MK, Thompson EJ. In vivo relationship of tumor necrosis factor-alpha to blood-brain barrier damage in patients with active multiple sclerosis. J Neuroimmunol 1992; 38:27-33.
22. Selmaj K, Raine CS, Cannella B et al. Identification of lymphotoxin and tumor necrosis factor in multiple sclerosis lesions. J Clin Invest 1991; 87:949-954.
23. Tsukada N, Matsuda M, Miyagi K et al. Increased levels of intercellular adhesion molecule-1 (ICAM-1) and tumor necrosis factor receptor in the cerebrospinal fluid of patients with multiple sclerosis. Neurology 1993; 43:2679-2682.
24. Adachi K, Kumamoto T, Araki S. Elevated soluble interleukin-2 receptor levels in patients with active multiple sclerosis. Ann Neurol 1990; 28:687-691.
25. Matsuda M, Tsukada N, Miyagi K et al. Increased levels of soluble tumor necrosis factor receptor in patients with multiple sclerosis and HTLV-1-associated myelopathy. J Neuroimmunol 1994; 52:33-40.
26. Renno T, Krakowski M, Piccirillo C et al. TNF-alpha expression by resident microglia and infiltrating leukocytes in the central nervous system of mice with experimental allergic encephalomyelitis. Regulation by Th1 cytokines. J Immunol 1995; 154:944-953.
27. Martin R, Howell MD, Jaraquemada D et al. A myelin basic protein peptide is recognized by cytotoxic T cells in the context of four HLA-DR types associated with multiple sclerosis. J Exp Med 1991; 173:19-24.
28. Zhang J, Markovic-Plese S, Lacet B et al. Increased frequency of interleukin 2-responsive T cells specific for myelin basic protein and proteolipid protein in peripheral blood and cerebrospinal fluid of patients with multiple sclerosis. J Exp Med 1994; 179:973-984.
29. Correale J, Gilmore W, McMillan M et al. Patterns of cytokine secretion by autoreactive proteolipid protein-specific T cell clones during the course of multiple sclerosis. J Immunol 1995; 154:2959-2968.

30. Ben-Nun A, Wekerle H, Cohen IR. The rapid isolation of clonable antigen-specific T lymphocyte lines capable of mediating autoimmune encephalomyelitis. Eur J Immunol 1981; 11:195-199.
31. Ben-Nun A, Lando Z. Detection of autoimmune cells proliferating to myelin basic protein and selection of T cell lines that mediate experimental autoimmune encephalomyelitis (EAE) in mice. J Immunol 1983; 130:1205-1209.
32. Ando DG, Clayton J, Kono D et al. Encephalitogenic T cells in the B10.PL model of experimental allergic encephalomyelitis (EAE) are of the Th-1 lymphokine subtype. Cell Immunol 1989; 124:132-143.
33. Chen Y, Kuchroo VK, Inobe J et al. Regulatory T cell clones induced by oral tolerance: suppression of autoimmune encephalomyelitis. Science 1994; 265:1237-1240.
34. Powell MB, Mitchell D, Lederman J et al. Lymphotoxin and tumor necrosis factor-alpha production by myelin basic protein-specific T cell clones correlates with encephalitogenicity. Int Immunol 1990; 2:539-544.
35. Racke MK, Bonomo A, Scott DE et al. Cytokine-induced immune deviation as a therapy for inflammatory autoimmune disease. J Exp Med 1994; 180:1961-1966.
36. Santambrogio L, Crisi GM, Leu J et al. Tolerogenic forms of auto-antigens and cytokines in the induction of resistance to experimental allergic encephalomyelitis. J Neuroimmunol 1995; 58:211-222.
37. Jacobs CA, Baker PE, Roux ER et al. Experimental autoimmune encephalomyelitis is exacerbated by IL-1 alpha and suppressed by soluble IL-1 receptor. J Immunol 1991; 146:2983-2989.
38. Rott O, Fleischer B, Cash E. Interleukin-10 prevents experimental allergic encephalomyelitis in rats. Eur J Immunol 1994; 24:1434-1440.
39. Cash E, Minty A, Ferrara P et al. Macrophage-inactivating IL-13 suppresses experimental autoimmune encephalomyelitis in rats. J Immunol 1994; 153:4258-4267.
40. Camenga DL, Johnson KP, Alter M et al. Systemic recombinant alpha-2 interferon therapy in relapsing multiple sclerosis. Arch Neurol 1986; 43:1239-1246.
41. Knobler RL, Panitch HS, Braheny SL et al. Systemic alpha-interferon therapy of multiple sclerosis. Neurology 1984; 34:1273-1279.
42. Durelli L, Bongioanni MR, Cavallo R et al. Chronic systemic high-dose recombinant interferon alpha-2a reduces exacerbation rate, MRI signs of disease activity, and lymphocyte interferon gamma production in relapsing-remitting multiple sclerosis. Neurology 1994; 44:406-413.
43. Jacobs L, Salazar AM, Herndon R et al. Multicentre double-blind study of effect of intrathecally administered natural human fibroblast interferon on exacerbations of multiple sclerosis. Lancet 1986; 2:1411-1413.
44. Milanese C, Salmaggi A, La Mantia L et al. Double blind study of intrathecal beta-interferon in multiple sclerosis: clinical and laboratory results. J Neurol Neurosurg Psychiatry 1990; 53:554-557.

45. Paty DW, Li DK, the UBC MS/MRI Study Group et al. Interferon beta-1b is effective in relapsing-remitting multiple sclerosis. II. MRI analysis results of a multicenter, randomized, double-blind, placebo-controlled trial. Neurology 1993; 43:662-667.
46. The IFNB Multiple Sclerosis Study Group. Interferon beta-1b is effective in relapsing-remitting multiple sclerosis. I. Clinical results of a multicenter, randomized, double-blind, placebo- controlled trial. Neurology 1993; 43:655-661.
47. The IFNB Multiple Sclerosis Study Group and the University of British Columbia MS/MRI Analysis Group. Interferon beta-1b in the treatment of multiple sclerosis: final outcome of the randomized controlled trial. Neurology 1995; 45:1277-1285.
48. Voorthuis JA, Uitdehaag BM, De Groot CJ et al. Suppression of experimental allergic encephalomyelitis by intraventricular administration of interferon-gamma in Lewis rats. Clin Exp Immunol 1990; 81:183-188.
49. Kuroda Y, Shimamoto Y. Human tumor necrosis factor-alpha augments experimental allergic encephalomyelitis in rats. J Neuroimmunol 1991; 34:159-164.
50. Racke MK, Sriram S, Carlino J et al. Long-term treatment of chronic relapsing experimental allergic encephalomyelitis by transforming growth factor-beta 2. J Neuroimmunol 1993; 46:175-183.
51. Santambrogio L, Hochwald GM, Saxena B et al. Studies on the mechanisms by which transforming growth factor-beta (TGF-beta) protects against allergic encephalomyelitis. Antagonism between TGF-beta and tumor necrosis factor. J Immunol 1993; 151:1116-1127.
52. Kuruvilla AP, Shah R, Hochwald GM et al. Protective effect of transforming growth factor beta 1 on experimental autoimmune diseases in mice. Proc Natl Acad Sci USA 1991; 88:2918-2921.
53. Zurawski G, de Vries JE. Interleukin 13, an interleukin 4-like cytokine that acts on monocytes and B cells, but not on T cells. Immunol Today 1994; 15:19-26.
54. Link J, He B, Navikas V et al. Transforming growth factor-beta 1 suppresses autoantigen-induced expression of pro-inflammatory cytokines but not of interleukin-10 in multiple sclerosis and myasthenia gravis. J Neuroimmunol 1995; 58:21-35.
55. Duong TT, St. Louis J, Gilbert JJ et al. Effect of anti-interferon-gamma and anti-interleukin-2 monoclonal antibody treatment on the development of actively and passively induced experimental allergic encephalomyelitis in the SJL/J mouse. J Neuroimmunol 1992; 36:105-115.
56. Billiau A, Heremans H, Vandekerckhove F et al. Enhancement of experimental allergic encephalomyelitis in mice by antibodies against IFN-gamma. J Immunol 1988; 140:1506-1510.
57. D'Andrea A, Rengaraju M, Valiante NM et al. Production of natural killer cell stimulatory factor (interleukin 12) by peripheral blood mononuclear cells. J Exp Med 1992; 176:1387-1398.

CHAPTER 8

Towards Therapeutic Manipulation of Th1 and Th2 Cell Responses

Luciano Adorini

INTRODUCTION

Following the initial description of two subsets of differentiated $CD4^+$ T lymphocytes in the mouse,[1] analysis of T cell clones from allergic patients clearly defined antigen-specific Th1 and Th2 cells in humans.[2] As doubts were dispelled,[3] the Th1/Th2 paradigm rapidly became a key concept in immunoregulation.

Mouse and human $CD4^+$ as well as $CD8^+$[4] T cells can be distinguished, based on their pattern of cytokine production, into three major subsets: Th1, Th2, and Th0. The distinction between type 1 and type 2 may indeed represent a fundamental dichotomy for all T cell subsets.[5] Th1 cells are characterized by secretion of interferon-γ (IFN-γ), IL-2, and TNF-β, and they promote cell-mediated immunity and the ability to eliminate intracellular pathogens. Conversely, Th2 cells selectively produce IL-4 and IL-5, and are involved in the development of humoral immunity protection against extracellular pathogens. Th0 cells, which could either represent precursors of Th1/Th2 cells or a terminally differentiated subset, are not restricted in their lymphokine production. The development of Th1 and Th2 cells is influenced by several factors, of which two are most important: cytokines and ligand-TCR interaction. Among the cytokines, decisive roles are played by IL-12 and IL-4, guiding T cell responses towards the Th1 or Th2 phenotype, respectively.[6,7]

Polarized Th1 and Th2 subsets can be generated from $CD4^+$ populations in vitro,[8] recovered from primed animals,[9] and found in patients suffering from autoimmune or allergic diseases.[10] However, polarized Th1 and Th2 cells represent extremes in a spectrum. Detection of intracytoplasmic cytokine production by polarized Th1 and Th2 cell populations analyzed at the single-cell level

Immunointervention in Autoimmunity by Th1/Th2 Regulation, edited by Luciano Adorini.

has confirmed the existence of defined Th1 and Th2 cells, selectively producing IFN-γ or IL-4, respectively, but has also revealed intermediate patterns.[11] Within this spectrum, discrete subsets of differentiated T cells secreting a mixture of Th1 and Th2 cytokines may exist, for example mouse T cells secreting IFN-γ and IL-10.[12] It has been suggested that the polarization of the immune response may be achieved by altering the probabilities of IFN-γ, IL-4 and other cytokine gene expression at the population level, rather than by selective expansion of distinct T cell subsets.[13,14] However, molecular mechanisms to explain the polarization of Th1 and Th2 subsets, based on the differential expression of the receptors for IFN-γ and IL-12, do exist. The ability of IFN-γ to inhibit the proliferation of Th2 but not of Th1 cells may be related to lack of IFN-γR β chain expression in Th1 cells.[15] IFN-γR β chain loss also occurs in IFN-γ-treated Th2 cells, and therefore does not represent a Th1 cell-specific differentiation event.[16] Conversely, developmental commitment to the Th2 lineage results from rapid loss of IL-12 signaling in Th2 cells.[17] The inability of Th2 cells to respond to IL-12 may be due to downregulation of IL-12R β2 subunit.[18] These findings are therefore consistent with a general model in which selective modulation of cytokine signaling plays an important role in the acquisition of distinct Th cell phenotypes.

The reciprocal regulation between Th cell subsets is another driving force polarizing $CD4^+$ T cells into differentiated Th1 or Th2 cells. IL-12 promotes the development of Th1 cells[8,19,20] and inhibits IL-4-induced IgE synthesis.[21] IFN-γ amplifies the IL-12-dependent development of Th1 cells[22] and inhibits Th2 cell proliferation.[23] Conversely, IL-4 and IL-10 inhibit lymphokine production by Th1 clones.[24] In addition, IL-10,[25] IL-4 and IL-13[26] suppress the development of Th1 cells through downregulation of IL-12 production by monocytes.

Th1 and Th2 Cells in Autoimmune Diseases

Experimental models

Th1 cells are considered to be involved in the induction of experimental autoimmune diseases.[27-30] Evidence for this is based on adoptive transfer experiments demonstrating that $CD4^+$ cells producing Th1-type lymphokines can transfer disease, both in experimental allergic encephalomyelitis (EAE)[31] and in insulin dependent diabetes mellitus (IDDM)[32-34] models. However, cytokine regulation is complex, for example TNF-α and IL-10 have opposite effects on IDDM depending on the developmental stage of the immune system.[35,36] This could also explain why, in some cases, β cell destruction in IDDM has been associated with Th2 rather than Th1 cells.[37,38]

The reciprocal regulation between T cell subsets predicts a role for Th2 cells in inhibition of autoimmune diseases. Regulatory T cells that suppress the development of EAE produce Th2-type cytokines,[39] and recovery from EAE is associated with increased Th2 cytokines in the CNS.[40] In addition, administration of IL-4 to mice with EAE ameliorates the disease.[41] These results clearly suggest that activation of Th2 cells may prevent EAE. Evidence for a protective role of Th2 cells is also provided by the reduced IDDM incidence following IL-4[42] or IL-10[43] administration to nonobese diabetic (NOD) mice. A role for Th2 cells regulating the onset of IDDM is also suggested by their capacity to inhibit the spontaneous onset of diabetes in rats[44] and by the correlation between protection from IDDM and IL-4 production in double-transgenic mice on BALB/c background.[45] Transgenic NOD mice that express IL-4 in their pancreatic β cells are protected from insulitis and IDDM, a direct indication that Th2 cytokines can prevent destructive autoimmunity.[46]

However, Th2 cells transgenic for a TCR derived from a clone able to transfer IDDM, when injected into neonatal NOD mice invaded the islets but did not provoke disease and neither did they provide substantial protection.[34] Similar results were also obtained by adoptive transfer of nontransgenic Th1 and Th2 cell lines into neonatal mice.[47] Therefore, these data do not support the concept that Th2 cells afford protection from IDDM, at least in the effector phase of the disease. Rather, they are in accord with the observation that transgenic expression by islet cells of IL-10,[36,48] an inhibitory lymphokine of Th1 cells, actually promotes insulitis and IDDM.

Collectively, these results point to a critical role of Th1 cells in the induction of autoimmune diseases, whereas the influence of Th2 cells is still unclear. In any case, whether or not Th2 cells exert a direct protective role, diversion away from proinflammatory Th1 cells should be effective in reducing the chronic inflammatory response which is typical of autoimmune diseases.

Human diseases

Th1 cells appear to be involved also in human organ-specific autoimmune diseases. $CD4^+$ T cell clones isolated from lymphocytic infiltrates of Hashimoto's thyroiditis or Graves' disease exhibit a clear-cut Th1 phenotype.[49] In addition, most T cell clones derived from peripheral blood or cerebrospinal fluid of multiple sclerosis patients show a Th1 lymphokine profile.[50]

Involvement of Th1 cells has also been suggested in other human autoimmune diseases. Insulitis in IDDM patients has been shown to comprise a large number of IFN-γ-producing lymphocytes.[51] T cell clones derived from the synovial membrane of rheumatoid arthritis (RA) patients also display a Th1 phenotype as they produce, upon activation, large amounts of IFN-γ and no or very little IL-4.[52] Another study has shown that most $CD4^+$ and $CD8^+$ clones recovered from synovial fluid of RA patients display a Th1 phenotype.[53] Interestingly, in situ hybridization for T cell cytokine expression demonstrates a Th1-like pattern in most synovial samples from RA patients, whereas samples from patients with reactive arthritis, a disorder with similar synovial pathology but driven by persisting exogenous antigen, express a Th0 phenotype.[54]

The situation is less clear in most systemic autoimmune disorders. In general, heterogeneous cytokine profiles are found in the serum or target organs of patients with systemic autoimmunity, such as systemic lupus, Sjögren's syndrome, and primary vasculitis.[10]

IMMUNOINTERVENTION IN AUTOIMMUNE DISEASES BY ALTERING THE TH1/TH2 BALANCE

The emerging picture of different Th subset development following different modes or doses of antigen administration allows the integration of the concept of distinct functional T cell subsets into some of the mechanisms of tolerance, including deletion, anergy and suppressor T cells.

In addition, the Th1/Th2 concept has emphasized a mechanism of immunoregulation that could explain some forms of tolerance: immune deviation.[55] A possible example of immune deviation is provided by autoantigen-specific therapy of IDDM in NOD mice. Autoantigens in IDDM are fairly well known.[56] Among them, glutamic acid decarboxylase (GAD) appears to be most important, because responses to GAD are detected before responses to other autoantigens, including insulin, heat shock protein, peripherin and carboxypeptidase H.[57,58] Intravenous or intrathymic administration of GAD to three week-old NOD mice, which do not yet display islet infiltration, prevents T cell proliferation to GAD in 12 week-old mice, and also prevents the development of intraislet infiltration and IDDM in adult NOD mice. The reduction in the number of IFN-γ-secreting GAD-specific T cells[57] associated with the continuing production of autoantibodies to GAD[58] suggests that parenteral GAD administration has indeed induced a biased switch towards a Th2 response. Evidence for induction of a protective Th2 response has recently been obtained by nasal administration of GAD peptides to NOD mice.[59] These data indicate that nasal administration of GAD65 peptides induces a Th2 cell response that inhibits the spontaneous development of autoreactive Th1 responses and the progression of β cell autoimmunity in NOD mice.

Additional approaches to immune deviation are emerging, such as DNA vaccination. The TCR variable chain Vβ8.2 is expressed on pathogenic T cells that induce EAE in $H\text{-}2^u$ mice immunized with myelin basic protein (MBP).[60] Vaccination of these mice with naked DNA encoding Vβ8.2 protects them from EAE, and protection has been found associated to a shift from a Th1 to a Th2 response.[61]

Collectively, these results suggest that immune deviation towards the Th2 phenotype may be effective in treating Th1-mediated autoimmune diseases. Essentially, two types of strategies are available: cytokine-based and antigen-based (Table 8.1). A critical point for their clinical applicability of this approach is the possibility to modify the Th1/Th2 balance in antigen-primed individuals. This appears to be the case,[62] at least for induction of unresponsiveness in Th1-type cells, therefore it may be possible to treat established autoimmune diseases.

CYTOKINE-BASED IMMUNOINTERVENTION

Cytokines are essential components of the immune response, and an imbalance in the cytokine network plays an important role in the initiation and perpetuation of autoimmune diseases. Elucidation of the structure and function of individual cytokines and cytokine receptors, as well as their selective expression by Th1 and Th2 $CD4^+$ T cells, are already contributing to the identification of novel strategies for selective immunosuppression.

Several approaches can be used to inhibit a given cytokine: the cytokine itself can be neutralized by specific monoclonal antibodies or by soluble cytokine receptors, whereas the cytokine receptor can be inhibited by monoclonal antibodies or by receptor antagonists competing with the ligand for the receptor binding site (see Table 8.1).

Cytokine-based manipulation of Th1 and Th2 responses, altering the balance between the two differentiated Th subsets, offers a unique possibility to interfere with autoimmune diseases, namely immune deviation from a proinflammatory to an anti-inflammatory response.

IL-12, a Th1-inducing cytokine

IL-12 is a heterodimer composed of two covalently linked glycosylated chains, p35 and p40, encoded by distinct genes.[63,64] This cytokine, produced predominantly by activated monocytes and dendritic cells but also by other cell types such as keratinocytes, enhances proliferation and cytolytic activity of NK and T cells, and stimulates their IFN-γ production.[65] Most importantly, IL-12 induces the development of Th1 cells in vitro[8,19] and in vivo.[20] In addition, IL-12 is a potent cofactor stimulating growth, IFN-γ synthesis, and cell adhesion of already differentiated Th1 cells.[66] Several recent reports, summarized below, document the key role of IL-12 in the induction of Th1 cell-mediated autoimmune diseases.

Administration of IL-12 induces rapid onset of IDDM in 100% of NOD female mice, whereas only about 60-70% of control littermates eventually develop IDDM.[67] This effect is not due to toxicity of IL-12 for pancreatic β cells, as shown by the normal appearance of islet cells and by the absence of IDDM in BALB/c mice treated with IL-12. Acceleration of IDDM in genetically susceptible NOD mice is accompanied by increased Th1 cytokine production by islet-infiltrating $CD4^+$ and $CD8^+$ T cells, and by selective destruction of islet β cells,

Table 8.1. Strategies targeting Th1 cells in autoimmune diseases

Cytokine-based
- **Inhibition of IL-12 and/or IFN-γ**
  - → Anti-cytokine mAb or cytokine antagonists
  - → Anti-cytokine receptor mAb
  - → Soluble cytokine receptors
  - → Inhibitors of cytokine production
- **Administration of IL-4 and/or IL-10**

Antigen-based
- **Direct inhibition of Th1 cells**
  - → Systemic administration of soluble autoantigen
- **Induction of Th2 cells/deviation from Th1 cells**
  - → Oral administration or inhalation of autoantigen
  - → Presentation of autoantigen by B cells
  - → Administration of peptide analogues of autoantigen

Combined cytokine- and antigen-based
- → Administration of autoantigen together with antagonists or inhibitors of IL-12
- → Administration of autoantigen together with IL-4

suggesting a causal link between IL-12, Th1 cell induction, and development of IDDM.

Conversely, following a protocol developed by O'Hara and co-workers,[68] we could confirm that intermittent administration of IL-12 (once weekly for 12 weeks) to NOD mice delays and reduces IDDM development (S Trembleau et al, unpublished). Data explaining these opposite effects of IL-12 are not yet available. However, it is conceivable that intermittent administration of IL-12, while still favoring Th1 induction, may be unable to sustain Th1 cell development. Thus, the aborted induction of a Th1 response, possibly coupled to the emergence of regulatory Th2 cells, may result in delay of Th1-mediated disease progression. This result should not be considered a surprising paradox, but rather an expected property of Th1/Th2 cell regulation.

To determine the role of endogenous IL-12 in IDDM development, mice deficient in IL-12 were generated by targeted disruption of the gene encoding the p40 subunit[69] and backcrossed to the NOD background. Spleen cells from $IL\text{-}12^{-/-}$ NOD mice do not produce IL-12 p40 nor IL-12 p75 when stimulated with LPS or SAC and IFN-γ, and fail to produce IFN-γ after stimulation with SAC or LPS. Conversely, spleen cells from $IL\text{-}12^{-/-}$ NOD mice secrete IFN-γ when stimulated with plate-bound anti-TCR mAb, indicating that T cells from IL-$12^{-/-}$ mice have the capacity to produce IFN-γ. Confirming the results obtained in vitro, IL-$12^{-/-}$ NOD mice, after LPS administration, do not produce detectable IL-12 p40 nor IL-12 p75, and show a 95% reduction of serum IFN-γ levels. Injection of hen egg-white lysozyme (HEL) in IL-$12^{-/-}$ NOD mice induces antigen-specific T cells able to secrete IL-2, IL-4, IL-5 and IL-10 in amounts comparable to littermate controls, but are defective in IFN-γ secretion. Histological analysis of pancreata from 12 week-old IL-$12^{-/-}$ NOD mice shows almost complete absence of insulitis, whereas it is quite florid in IL-$12^{+/+}$ littermate controls, demonstrating the critical role of IL-12 in the generation of islet infiltrate. Preliminary results indicate a reduced incidence of spontaneous IDDM in IL-$12^{-/-}$ NOD mice as compared to controls. Moreover, IDDM development in a model of cyclophosphamide-accelerated diabetes is considerably reduced in IL-$12^{-/-}$ NOD mice as compared to littermate controls, indicating that endogenous IL-12 is required for IDDM development (S. Trembleau et al, manuscript in preparation). These results indicate the critical role of endogenous IL-12 in the induction of antigen-specific IFN-γ-producing cells, insulitis, and IDDM.

In a model of adoptively transferred EAE, mice injected with lymph node cells (LNC) stimulated in vitro with proteolipid protein (PLP) and IL-12 exhibit a more severe and prolonged disease than mice given LNC stimulated with PLP alone. Similarly, mice treated with IL-12 in vivo following the transfer of PLP-stimulated LNC develop a more severe and prolonged course of EAE, as compared to vehicle-treated controls. Most importantly, administration of anti-IL-12 antibodies substantially reduces the incidence and severity of adoptively transferred EAE, suggesting that endogenous IL-12 plays a key role in its pathogenesis.[70]

Treatment with IL-12 also enhances the autoimmune response of DBA/1 mice to type II collagen resulting in severe, destructive arthritis. IFN-γ production by collagen-specific $CD4^+$ T cells as well as the synthesis of complement-fixing antibodies of IgG2a and IgG2b isotypes is strongly upregulated, suggesting that IL-12-induced Th1 cells may have a crucial role in the pathogenesis of this form of arthritis.[71] Surprisingly, injection of high doses of IL-12 in DBA/1 mice immunized with collagen and mycobacteria ameliorates disease, suggesting that some initial events in the induction of collagen-induced arthritis can be suppressed by high doses of IL-12.[72]

Experimental colitis, a Th1-mediated autoimmune disease, can be treated by administration of anti-IL-12 antibodies even late after disease induction, suggesting that endogenous IL-12 may be required not only for induction but also for progression of the disease.[73]

Collectively, these results indicate that IL-12 may have a primary role in the induction of autoimmune diseases.[30] Given the critical role of IL-12 in the induction of autoimmune diseases, IL-12 antagonists become very attractive candidates for immunointervention. Besides anti-IL-12 antibodies, other possible antagonists include antibodies to the IL-12 receptor (IL-12R) or soluble IL-12R-IgG fusion constructs, based on the recently cloned IL-12R subunit.[74,75] IL-12 is not only a differentiation factor essential for the development of Th1 cells, but also a costimulus for activation of effector Th1 cells,[66,76] suggesting that administration of IL-12 antagonists may represent a strategy not only to prevent but possibly also to treat Th1-mediated autoimmune diseases.

IL-10, a Th1-inhibiting cytokine

IL-10 is a potent suppressor of several effector functions of macrophages, T cells and NK cells. In addition, it contributes to regulate proliferation and differentiation of B cells, mast cells, and thymocytes.[24] Addition of IL-10 during macrophage activation selectively inhibits up-regulation of B7 expression, a costimulatory molecule critical for T cell activation.[77] However, the most important property of IL-10 for manipulating the Th1/Th2 cell balance, is its capacity to inhibit Th1 cells. The inhibition of the Th1 cell pathway by IL-10 is mediated by several mechanisms, including inhibition of IL-12 production by APC[25] and blocking of IFN-γ synthesis by differentiated Th1 cells.[24] Thus, IL-10 indirectly favors Th2 cell development by inhibiting Th1 cell induction.

Production of pro-inflammatory monokines such as IL-1, IL-6, IL-8, TNF-α and GM-CSF following activation of human or mouse macrophages is strongly inhibited by IL-10.[78,79] Production of reactive oxygen and nitrogen species by IFN-γ activated macrophages is also inhibited by IL-10.[80] These results indicate that IL-10 has strong anti-inflammatory properties. In addition, IL-10 induces a long-term antigen-specific anergic state in human $CD4^+$ T cells, and therefore may also play an important role in the induction and maintenance of antigen-specific T cell tolerance.[81]

IL-10 injection in adult NOD mice has been shown to decrease insulitis and IDDM,[43] indicating that systemic administration of this cytokine may affect the course of autoimmune diseases. However, this may not be true in every situation, as demonstrated by the observation that transgenic IL-10 expression under the insulin promoter does not prevent islet destruction and may actually enhance the inflammatory response.[36,48] The discrepancy between the protective effect of IL-10 administration to adult NOD mice[43] versus the precipitating effect of transgenic IL-10 expression in pancreatic islet α[48] or β[36] cells points to the importance of whether cytokines are produced locally or given systemically. Alternatively, the different effects of IL-10 on IDDM induction could be related to the developmental stage of the immune system or of the target cell at the time of cytokine delivery.

It is likely that these inhibitory functions of IL-10 could be exploited clinically to prevent rejection and prolong graft survival.[82] Its activity in inhibiting macrophage activation and Th1 cytokine synthesis also suggests a possible use as a nonantigen-specific suppressor factor in the treatment of autoimmune diseases. This possibility would be in accord with the observation that mice failing to make IL-10 because of targeted disruption of its gene develop a severe inflammatory bowel disease.[83] The disease may reflect an overstimulation of Th1 cells, not controlled by Th2-derived IL-10, to gut antigenic stimulation. Supporting this hypothesis, severe colitis was abrogated in a model of inflammatory bowel disease by systemic administration of IL-10 but, interestingly, not of IL-4.[84]

IL-4, a Th2-inducing cytokine

IL-4 is the cytokine playing the most important role in Th2 cell development. In both polyclonal and antigen-specific T cell responses, IL-4 promotes development of Th2 cells while inhibiting development of Th1 cells. Using TCR transgenic T cells, it has been shown that IL-4 drives the development of Th2 cells, and this effect is dominant over that of IL-12.[8] IL-4 promotes development of Th2 cells against *Leishmania major*[85] and anti-IL-4 antibody treatment can render a susceptible mouse strain resistant to the parasite.[86] IL-4 and IL-10 synergize to inhibit the development of delayed-type hypersensitivity to *Leishmania major* in *Leishmania*-infected mice.[87] An important function of Th2 cells, and hence of IL-4, appears thus to be the control of the tissue-damaging effects of proinflammatory cytokines secreted or induced by Th1 cells.

Based on these results, treatment of experimental autoimmune diseases by IL-4 has been attempted. Transgenic NOD mice that express IL-4 in their pancreatic β cells are protected from insulitis and IDDM, indicating the feasibility of a peripheral approach to the treatment of autoimmunity.[46] IL-4 administration to adult NOD mice inhibits IDDM development, although insulitis is not blocked.[42] This would be in accord with the hypothesis that islet infiltration is not necessarily associated to IDDM, if Th2-type cells predominate.[88] For example, in NOD male mice, which develop insulitis but only a low incidence of IDDM, IL-4 mRNA expression is associated with nondestructive insulitis.[89] In streptococcal-induced arthritis IL-4 administration suppresses the chronic destructive phase[90] and in EAE it inhibits considerably the clinical manifestations of the disease.[41] Although these results look promising, systemic administration of IL-4 in man leads to severe side effects, which prevent its use in chronic diseases like autoimmune diseases.

Antigen-Based Immunointervention

Antigen itself can be used to selectively influence the development of antigen specific Th1 and Th2 cells. In this respect, the two most important variables are dose and route of administration. In addition, altered peptide ligands can modify the profile of lymphokine production by T cells, thus affecting the Th1/Th2 phenotype.

Dose

The dose of antigen can influence whether a cell-mediated or humoral class of immune response is elicited, and this may be largely accounted for by the development of Th populations producing distinct sets of cytokines. In vitro priming of TCR-transgenic $CD4^+$ T cells with high antigen doses leads to selective differentiation into Th1 cells, whereas low doses of the same peptide antigen induce cells with the same transgenic TCR to differentiate into Th2 cells.[91] High and low antigen doses are, however, relative, and in another TCR transgenic model intermediate doses of peptide antigen in vitro gave rise to Th1 cells, whereas Th2 cells preferentially developed at very high or very low antigen doses.[92] Continuous administration to BALB/c mice of any soluble protein antigen tested induces a long-lasting antigen-specific unresponsiveness of $CD4^+$ Th1 cells and a vigorous priming of Th2 cells.[93] In this model, continuous administration of low rather than high doses of antigen is most efficient in Th2 cell induction. Thus, antigen dose is an important factor controlling the differentiation fate of a naive $CD4^+$ cell in vivo. In different models of autoimmune diseases, transmucosal administration of antigen at low doses has been shown to favor induction of Th2 cells and to inhibit disease.[94]

Route of administration

Parenteral antigen

Parenteral administration of soluble antigen selectively tolerizes primarily Th1 but not Th2-type cells.[95,96] We have recently shown that continuous administration of low dose soluble antigen to BALB/c mice inhibits Th1 while actually stimulating Th2 cell development, as revealed by decreased production of IL-2 and IFN-γ and by a strong increase in the secretion of IL-4 and IL-5 by antigen-specific $CD4^+$ T cells. Inhibition of Th1 cells can be induced in any mouse strain by administration of soluble antigen. Conversely, selective induction of Th2 cells by soluble antigen administration is controlled by non-MHC gene polymorphism: BALB/c mice readily develop Th2 cells whereas B10.D2 do not.[93]

The default development of Th2 cells in BALB/c could depend on the initial activation of IL-4-producing cells such as the $CD4^+NK1.1^+$ population.[97]

This is unlikely, because a similar induction of Th2 cells was obtained in normal BALB/c and in β2M-deficient BALB/c mice,[93] which lack this cell population whose development requires MHC class I expression.[98,99] Rather, the IL-4 required for Th2-priming in vivo could be produced by antigen-specific class II-restricted $CD4^+$ T cells themselves. Indeed, in susceptible mouse strains, *L. major* infection induces very rapid IL-4 production by $CD4^+NK1.1^-$ cells.[100]

The APC type presenting soluble antigen may also play a role in selectively priming Th2 cells. B cells appear to be the crucial APC for the development of Th2 responses.[101] Induction of EAE, a Th1-mediated autoimmune disease, has been prevented by targeting the autoantigen to B cells,[102] and this prevention has been recently found to be associated with the priming of antigen-specific Th2 cells.[103] Dendritic cells and macrophages, but not B cells, secrete IL-12 in vivo (Guéry et al, submitted). Therefore, antigen presentation by APCs which do not produce IL-12, such as B cells, could favor Th2 development. IgD targeting on the B cell surface by bivalent antibody fragments results in B cell activation,[104] which might be a prerequisite for priming of Th2 cells.[103] If this is the case, targeting antigen to B cells might be a way to force the immune system towards a Th2-type immune response, an approach potentially applicable to the treatment of autoimmune diseases.

Interestingly, soluble protein but not peptide administration diverts the immune response of a clonal $CD4^+$ T cell population to the Th2 cell pathway.[105] This aspect of immune deviation has been examined in vivo by transferring naive TCR transgenic T cells specific for the chicken OVA peptide 323-339 presented by $I\text{-}A^d$ molecules into normal BALB/c mice. Analysis of Th phenotype indicates that pretreatment with soluble OVA suppresses Th1 cell differentiation in favor of the generation of Th2 cells producing IL-4 and IL-5. Pretreatment with soluble peptide 323-339 also inhibits Th1 cell development, but fails to induce Th2 cell differentiation. Thus, pretreatment with soluble protein antigen or with synthetic peptide both inhibit Th1 cell development, but only protein and not peptide administration can deviate the in vivo response of a clonal T cell population from the Th1 to the Th2 pathway. A possible interpretation of this finding is that different types of APCs could preferentially present soluble protein and peptide. Although the relative role of dendritic cells and B cells in the presentation of soluble antigen in vivo is not yet completely clear, it appears that immature dendritic cells can present soluble protein[106] or peptide[107] to $CD4^+$ cells whereas B cells preferentially present protein antigen.[108,109] As antigen targeting to B cells favors Th2 cell differentiation it is possible that soluble OVA presented by B cells leads to commitment of naive $KJ1\text{-}26^+$ $CD4^+$ cells to the Th2 pathway. The difference between soluble protein and peptide in the capacity to induce Th2 cell responses could also be explained by their different half-lives in vivo. In this case, time of antigen exposure could significantly influence priming of Th2, but not inhibition of Th1 responses. Different persistence times are known to affect the potency and immunodominance of peptide-class II complexes recognized by T cells[110] and this could also play a role in the regulation of Th1/Th2 cell induction. Alternatively, processing of OVA in vivo could give rise to epitopes with different affinities for the DO11.10 TCR and low-affinity interactions could lead to induction of Th2 cells, in analogy to altered peptide ligands.[111-113]

In oral tolerance, induction of antigen-specific $CD8^+$ suppressor T cells was found to be an important mechanism for T cell unresponsiveness.[114,115] This is not the case following continuous subcutaneous administration of antigen, since a similar inhibition of Th1 cells was obtained in normal mice and mice lacking $CD8^+$ T cells. While induction of $CD8^+$ suppressor T cells can be definitively ruled out, immunoregulation by antigen-specific Th2 cells could play a role in our model. The polarization towards Th2 cells appears to be mediated by a dual mechanism: induction of Th1 cell unresponsiveness and default development of Th2 cells. The different antigen dose dependence of these two events suggests that, following continuous administration of soluble proteins, these two pathways may develop independently. This is further supported by experiments demonstrating that blocking endogenous IL-4 during continuous administration of soluble antigen inhibits Th2 cell development, but does not reverse the inhibition of Th1 cells.[93] Upon immunization with antigen in adjuvant, the primed Th2 population would then be preferentially expanded and could influence further differentiation of naive Th cells in situ by the secretion of IL-4 and IL-10.[116]

Oral tolerance

The Th1/Th2 balance can also be altered by oral administration of antigens. Oral administration of antigen at lower doses can apparently induce suppressor T cells that act by releasing inhibitory cytokines after antigen-specific stimulation, as demonstrated by adoptive transfer of suppression by $CD8^+$ T cells, which produce transforming growth factor β (TGF-β) following antigen-specific triggering.[115]

In addition to $CD8^+$ TGF-β secreting cells, $CD4^+$ regulatory cells are also induced.[117] These $CD4^+$ cells are of Th2 type, secreting IL-4 and IL-10 in addition to TGF-β. The secretion of Th2-type cytokines induced by a myelin antigen such as MBP could also inhibit pathogenic responses driven by other myelin antigens, like PLP, by a mechanism of bystander suppression.[94] Bystander suppression, which could be considered specular to epitope spreading,[118] could be instrumental to rendering antigen-based immunointervention feasible. Accordingly, the induction of a Th2 response to a self antigen present in the local microenvironment could convert the response to any autoantigen in the inflammatory site into a protective, or at least nonpathogenic, response.

Feeding higher antigen doses, both Th1 and Th2 cells are deleted by apoptosis after their initial activation, demonstrating that orally administered antigens can induce tolerance not only by active suppression, but also by extrathymic deletion of antigen-reactive cells.[119] Repeated intravenous administration of large amounts of soluble MBP could also delete peripheral autoreactive T cells, and thereby improve the course of EAE.[120] This implies that the stronger the initial proliferative response to antigen, the more vulnerable T cells are to apoptosis by TCR re-engagement. Deletion of pathogenic autoreactive T cells by infusion of high dose autoantigen has obvious therapeutic potential for the treatment of autoimmune diseases, although possible side effects should not be underestimated.

Suppression of EAE, either induced by oral tolerization or by natural recovery, could be related to the secretion of inhibitory cytokines like TGF-β and IL-4, actively suppressing the inflammatory process in the target organ.[40] These results suggest that gut-associated T cells may develop preferentially toward Th2, rather than Th1, phenotype. TGF-β can also enhance the antigen-dependent encephalitogenic function of T cells in EAE,[121] suggesting different modes of action of this lymphokine depending on the cell types present in a particular microenvironment. Accordingly, it has been reported that an encephalitogenic peptide can inhibit EAE when administered by inhalation, but not orally.[122]

In conclusion, experimental models of oral tolerance indicate that feeding low antigen doses can induce antigen-specific cells that produce TGF-β, IL-4 and IL-10. At higher antigen doses both Th1 and Th2 cells are deleted following their initial activation. Therefore, orally administered antigen can induce immunosuppression by activating Th2-type cells to secrete inhibitory cytokines, but also by extrathymic deletion of antigen-reactive Th1 and Th2 cells.[119]

Based on the results obtained in animal models, oral tolerance is currently being tested as a possible treatment for human autoimmune diseases. A double-blind pilot trial of oral tolerization with myelin antigens in MS patients has been completed.[123] Although the small number of patients treated does not allow conclusions about clinical efficacy, the results show a lower incidence of disease exacerbations in the myelin-fed group. A randomized, double-blind trial involving 60 patients with severe active rheumatoid arthritis has shown a statistically significant decrease in the number of swollen joints and tender joints in subjects fed chicken type II collagen for 3 months,[124] encouraging further research on clinical application of oral tolerance.

Altered peptide ligands

In addition to cytokines such as IL-12 and IL-4, several factors controlling the extent of TCR ligation, such as peptide affinity for MHC molecules, antigen-presenting cell (APC) type involved, ligand density and accessory molecules, contribute to the development of Th1 or Th2 cells. In general, peptide-class II complexes that interact strongly with the TCR favor induction of Th1 cells, whereas ligands binding with lower avidity favor induction of Th2 cells. This concept has been clearly defined using altered peptide ligands (APL, antigenic peptides modified from the wild-type at amino acid residues affecting interaction with the TCR), indicating that peptide-class II complexes interacting with high affinity with the TCR induce Th1 cells, whereas peptide-class II ligands interacting with lower affinity prime for Th2 cell responses.[112] Therefore, the TCR-ligand interaction can control the differentiation of naive $CD4^+$ cells into Th1 or Th2 cells, both in vitro[125] and in vivo.[112] Conversely, the T cell specificity itself does not determine the effector cell type, because TCR transgenic T cells can be induced to develop towards either the Th1 or Th2 phenotype in the presence of exogenous IL-12 or IL-4, respectively.[34,126,127]

The TCR can sense minor changes in the ligand and transduce signals inducing the T cell to respond with a gradation of effector functions. In this respect, the TCR has a pharmacological behavior similar to other receptor systems, in which agonists, partial agonists, and antagonists can be identified by modifying the ligand.[128] Two separate signals are required for T cell activation, phosphorylation of CD3ζ and CD3γδε chains,[129] and it has been hypothezised that partial agonists induce phosphorylation of the CD3γδε complex in the absence of phosphorylation of the ζ chain.[128] Support for qualitative differences in the signal delivered by APL is now provided by the observation that phosphorylation of the ζ chain without activation of the ZAP-70 kinase are linked to the antagonist effect.[130,131]

This partial T cell signaling could take different forms, like anergy induction, cytokine production without proliferation, or cytolysis without proliferation. Both Th1[132] and Th2[133] clones can be rendered anergic in vitro by APL delivering an incomplete signal, which allows cytokine production in the absence of clonal expansion. Failure of T cells specific for self epitopes to proliferate may be a highly efficient way to control autoimmune responses in vivo.

A major difference between TCR and MHC antagonists is their relative efficiency: TCR antagonists for either $CD4^+$[134] or $CD8^+$[135] T cells are about two orders of magnitude more efficient than MHC antagonists. This would render, in principle, APL acting as TCR antagonists attractive tools to induce selective immunosuppression. TCR antagonists have been shown to occur naturally in HIV[136] and hepatitis B[137] virus infections. Mutated viral epitopes behaving as TCR antagonists inhibit the capacity of CTL to destroy infected target cells, thus effectively protecting the cell harboring the mutant virus from elimination by CTL specific for the wild-type epitope.

The potent bioactivity of TCR antagonists may have therapeutic potential by altering the Th1/Th2 balance. It has been recently demonstrated that TCR antagonist peptides can inhibit EAE induced by a proteolipoprotein (PLP) epitope in SJL mice.[138] In addition, EAE can be reversed by administration of a soluble APL variant of an MBP epitope, which reduces IFN-γ and TNF-α production.[139,140] Immunization with an autoantigenic peptide altered at the primary TCR contact point protects mice from the development of EAE by inducing T cells that are cross-reactive with the native PLP peptide and secrete Th0 and Th2 cytokines.[111] The observation that anti-IL-4 antibody treatment abrogates protection induced by the APL and that transfer of APL-induced T cell lines protects mice from EAE, further supports the hypothesis that APL mediates protection by inducing regulatory Th2 cells in vivo. Modulation of cytokine patterns of MBP-specific human autoreactive T cell clones by a single amino acid substitution of their peptide ligand has also been demonstrated.[113] In this case the APL induced selective secretion of TGF-β, raising the possibility that modulation of T cell cytokine secretion may be important in regulating the response of potentially pathogenic human T cells specific for self antigens. Along these lines, inhibition of experimental autoimmune diseases following administration of peptide analogues of autoantigenic epitopes, as observed in EAE induced by myelin basic protein[141-143] and in adjuvant arthritis[143] models could also be re-interpreted as due to selective induction by APL of Th2 responses, leading to inhibition of pathogenic Th1 cells.

Theoretically, if the epitopes inducing pathogenic T cells could be identified, administration of TCR antagonists could prevent, or possibly even treat, autoimmune diseases. The development of this approach for clinical use would certainly be accelerated by the implementation of protocols able to induce some form of bystander immunosuppression, thereby converting a microenvironment dominated by proinflammatory Th1-type cytokines into one characterized by anti-inflammatory, Th2-type, cytokines. However, the feasibility of selective immunosuppression or immune deviation by APL acting as TCR antagonists requires detailed knowledge of the antigenic epitopes involved and homogeneity of these epitopes in different individuals. Obviously, the antagonist APL has to be designed in such a way that it will never become, under any circumstance, an agonist. These requirements may limit the clinical applicability of the APL approach, although this strategy is currently intensely explored.

PROSPECTS FOR TREATMENT OF HUMAN AUTOIMMUNE DISEASES

The results reviewed highlight the critical role of Th1 cells in the induction of several organ-specific autoimmune diseases and suggest a possible influence of Th2 cells in controlling, directly or indirectly, disease induction and/or progression. The efficacy of cytokine-specific treatments in chronic inflammatory disease states raises hopes for immunosuppressive strategies selectively targeting Th1 or Th2-type T cells. In particular, antagonists of IL-12 and inducers of

IL-4 or IL-10 offer the possibility to selectively manipulate Th1 and Th2 cell induction, with potential for the treatment of autoimmune diseases. The use of cytokine antagonists or cytokine inducers should in fact permit avoidance of systemic administration of cytokines and their severe side effects.

The considerable interest raised by cytokine-based immunotherapy is already being applied to clinical settings, particularly in rheumatoid arthritis (RA) patients.[144] For example, IL-10 administration is currently being tested in inflammatory bowel disease and in RA. As our understanding of Th1/Th2 regulation in human autoimmune diseases broadens, we will certainly witness the application of more articulate strategies able to selectively target cytokine production by Th1 or Th2 cells or to modify the Th1/Th2 balance in clinical situations (Table 8.1).

At present, consistent published results from clinical trials have reported targeting the proinflammatory cytokines IL-1 and TNF-α in RA patients, and results of TNF-α-directed therapies are particularly promising. Animal studies have clearly documented the important role of TNF-α in rheumatoid arthritis. Mice transgenic for the human TNF-α gene produce high levels of this cytokine and develop arthritis beginning at 4 weeks of age.[145] The disease is mediated by the transgenic molecule, as demonstrated by prevention of arthritis following administration of monoclonal antibodies to human TNF-α. In addition, in a model of type II collagen-induced arthritis, administration of anti-mouse TNF-α—even after disease onset—significantly reduced inflammation and tissue destruction.[146]

Based on these results, chimeric anti-TNF-α monoclonal antibody was administered to rheumatoid arthritis patients.[147] Treatment with anti-TNF-α was safe and well-tolerated, and led to significant clinical and laboratory improvements. After the first administration of anti-TNF-α antibody, remissions lasted, on the average, about three months, but reinjection of the antibody induced a significant anti-globulin response in most patients, reducing considerably the efficacy of the treatment. An alternative approach, using the soluble TNF receptor p55 chain fused to the constant region of human IgG1 heavy chain (sTNFR-IgG1), has been demonstrated to be about 10-fold more effective than anti-TNF-α antibody at neutralizing the activity of endogenous TNF, as assessed in a model of listeriosis[148] or in chronic relapsing EAE.[149] This fusion protein appears to achieve the same clinical effects as anti-TNF-α antibody administration without strong induction of neutralizing antibodies.

Even if clinical results of anti-TNF-α therapy in RA patients are very promising, the role of TNF-α in IDDM models is still puzzling. The fact that anti-TNF-α antibody treatment initiated before 3 weeks of age prevents insulitis and IDDM clearly suggests that TNF-α may be an essential mediator for the generation and/or activation of autoreactive lymphocytes.[35] Intriguingly, administration of TNF-α to adult NOD mice could also prevent IDDM, but the mechanism is still unclear.[35] TNF-α appears to have a distinct effect on the diabetogenic process depending upon the developmental stage of the immune system and of the target organ, perhaps in a manner analogous to IL-10. These results stress the importance of the time window of cytokine or anti-cytokine treatment to obtain the desired effect. If this concept cannot be translated to clinical practice, conditions to recreate a situation favoring the protective effects of the anti-cytokine treatment should be optimized.

Antigen-based immunointervention is also being tested clinically in autoimmune disease patients. Following the initial attempts to interfere with established human autoimmune diseases by oral administration of autoantigens,[123,124] clinical trials of oral tolerance to type II collagen in RA patients have been reported by two groups. Preliminary results[144] indicate some benefit only at the lowest dose tested in one of the two trials, but additional phase II and phase III trials are underway.

Although therapeutic manipulation of the Th1/Th2 balance in autoimmune disease patients by antigen administration is not yet proven, the apparent safety of the approach and the relatively straightforward putative mechanism of action encourage further research. The goal is to develop effective protocols capable to inhibit established pathogenic Th1 responses. In this respect antigen should be administered not simply to induce tolerance, but to convert the response in the lesion site from aggressive to protective, in short from Th1 to Th2. This should provide a suppressive environment for the response to any autoantigen and hopefully interfere with disease progression. A corollary of this hypothesis is that precise identification of the inciting autoantigen(s) is not a prerequisite for antigen-based immunointervention. Administration of any self antigen able to induce a strong Th2-type response locally should be sufficient to inhibit pathogenic Th1 cells. Clinical verification of this concept is eagerly awaited.

The most effective manipulation of Th1 and Th2 cells in autoimmunity may eventually rely on a combination of antigen and cytokine-based approaches. Ideally, they could be used to target specifically autoreactive T cells, diverting them from autoaggression by changing their lymphokine production profile. This strategy, which has been successfully applied to immunotherapy of parasitic diseases,[20,150] may complement attempts at treating human autoimmune diseases based on self antigen administration.

References

1. Mosmann TR, Cherwinski H, Bond MW et al. Two types of murine helper T cell clone. I Definition according to profile of lymphokine activities and secreted proteins. J Immunol 1986; 136:2348-57.

2. Del Prete G, De Carli M, Mastromauro C et al. Purified protein derivative of mycobacterium tuberculosis and excretory-secretory antigen(s) of toxocara canis expand in vitro human T cells with stable and opposite (type 1 T helper or type 2 T helper) profile of cytokine production. J Clin Invest 1991; 88:346-50.
3. Romagnani S. Human TH1 and TH2: doubt no more. Immunol Today 1991; 12:256-7.
4. Erard F, Wild M-T, Garcia-Sanz JA et al. Switch of CD8 T cells to noncytolytic $CD8^-CD4^-$ cells that make T_H2 cytokines and help B cells. Science 1993; 260:1802-5.
5. Carter LL, Dutton RW. Type 1 and type 2: a fundamental dichotomy for all T-cell subsets. Curr Opin Immunol 1996; 8:336-42.
6. Trinchieri G. Interleukin 12: a proinflammatory cytokine with immunoregulatory functions that bridge innate resistance and antigen-specific adaptive immunity. Annu Rev Immunol 1995; 13:251-76.
7. Paul WE, Seder RA. Lymphocytes responses and cytokines. Cell 1994; 76:241-51.
8. Hsieh C-S, Macatonia SE, Tripp CS et al. Development of Th1 $CD4^+$ T cells through IL-12 produced by *Listeria*-induced macrophages. Science 1993; 260:547-9.
9. Reiner SL, Locksley RM. The regulation of immunity to Leishmania major. Annu Rev Immunol 1995; 13:151-77.
10. Romagnani S. Lymphokine production by human T cells in disease states. Annu Rev Immunol 1994; 12:227-57.
11. Openshaw P, Murphy EE, Hosken NA et al. Heterogeneity of intracellular cytokine synthesis at the single-cell level in polarized T helper 1 and T helper 2 populations. J Exp Med 1995; 182:1357-67.
12. Mosmann TR, Sad S. The expanding universe of T-cell subsets: Th1, Th2 and more. Immunol Today 1996; 17:138-46.
13. Kelso A, Groves P, Troutt AB et al. Evidence for the stochastic acquisition of cytokine profile by $CD4^+$ T cells activated in a T helper type 2-like response in vivo. Eur J Immunol 1995; 25:1168-75.
14. Kelso A. Th1 and Th2 subsets: paradigms lost? Immunol Today 1995; 16:374-9.
15. Pernis A, Gupta S, Gollob KJ et al. Lack of interferon γ receptor β chain and the prevention of interferon γ signaling in Th1 cells. Science 1995; 269:245-7.
16. Bach EA, Szabo S, Dighe AS et al. Ligand-induced autoregulation of IFN-γ receptor β chain expression in T helper cell subsets. Science 1995; 270:1215-8.
17. Szabo SJ, Jacobson AG, Gubler U et al. Developmental commitment to the Th2 lineage by extinction of IL-12 signaling. Immunity 1995; 2:665-75.
18. Jacobson NG, Szabo SJ, Gueler ML et al. Regulation of interleukin-12 signal transduction during T helper phenotype development. Res Immunol 1995; 146:446-456.

19. Manetti R, Parronchi P, Giudizi MG et al. Natural killer cell stimulatory factor (interleukin 12, IL-12) induces T helper type 1 (Th1)-specific immune responses and inhibits the development of IL-4-producing Th cells. J Exp Med 1993; 177:1199-204.
20. Afonso LCC, Scharton TM, Vieira LQ et al. The adjuvant effect of interleukin-12 in a vaccine against *Leishmania major*. Science 1994; 263:235-7.
21. Kiniwa M, Gately M, Gubler U et al. Recombinant interleukin-12 suppresses the synthesis of IgE by interleukin-4 stimulated human lymphocytes. J Clin Invest 1992; 90:262-6.
22. Schmitt E, Hoehn P, Huels C et al. T helper type 1 development of naive $CD4^+$ T cells requires the coordinate action of interleukin-12 and interferon-γ and is inhibited by transforming growth factor-β. Eur J Immunol 1994; 24:793-8.
23. Gajewski TF, Fitch FW. Anti-proliferative effect of IFN-gamma in immune regulation. I. IFN-gamma inhibits the proliferation of Th2 but not Th1 murine helper T lymphocyte clones. J Immunol 1988; 140:4245-53.
24. Moore K, O'Garra A, de Waal Malefyt R et al. Interleukin-10. Annu Rev Immunol 1993; 11:165-90.
25. D' Andrea A, Aste-Amezaga M, Valiante NM et al. Interleukin 10 (IL-10) inhibits human lymphocyte interferon-γ production by suppressing natural killer cell stimulatory factor/IL-12 synthesis in accessory cells. J Exp Med 1993; 178:1041-8.
26. de Waal Malefyt R, Figdor CG, Huijbens R et al. Effects of IL-13 on phenotype, cytokine production, and cytotoxic function of human monocytes. Comparison with IL-4 and modulation by IFN-γ or IL-10. J Immunol 1993; 151:6370-81.
27. Powrie F, Coffmann RL. Cytokine regulation of T cell function: potential for therapeutic intervention. Immunol Today 1993; 14:270-4.
28. O'Garra A, Murphy K. T-cell subsets in autoimmunity. Curr Opin Immunol 1993; 5:880-6.
29. Liblau RS, Singer SM, McDevitt HO. Th1 and Th2 $CD4^+$ T cells in the pathogenesis of organ-specific autoimmune diseases. Immunol Today 1995; 16:34-8.
30. Trembleau S, Germann T, Gately MK et al. The role of IL-12 in the induction of organ-specific autoimmune diseases. Immunol Today 1995;
31. Ando DG, Clayton J, Kong D et al. Encephalitogenic T cells in the B10.PL model of experimental allergic encephalomyelitis (EAE) are of the Th1 lymphokine subtype. Cell Immunol 1989; 124:132-43.
32. Haskins K, McDuffie M. Acceleration of diabetes in young NOD mice with a $CD4^+$ islet specific T cell clone. Science 1990; 249:1433-6.
33. Bergman B, Haskins K. Islet-specific T-cell clones from the NOD mouse respond to beta-granule antigen. Diabetes 1994; 43:197-203.
34. Katz JD, Benoist C, Mathis D. T helper cell subsets in insulin-dependent diabetes. Science 1995; 268:1185-8.

35. Yang X-D, Tisch R, Singer SM et al. Effect of tumor necrosis factor α on insulin-dependent diabetes mellitus in NOD mice. I. The early development of autoimmunity and the diabetogenic process. J Exp Med 1994; 180:995-1004.
36. Wogesen L, Lee M-S, Sarvetnick N. Production of interleukin 10 by islet cells accelerates immune-mediated destruction of β cells in nonobese diabetic mice. J Exp Med 1994; 179:1379-84.
37. Anderson JT, Cornelius JG, Jarpe AJ et al. Insulin-dependent diabetes in the NOD mouse model. II. Beta cell destruction in autoimmune diabetes is a Th2 and not a Th1 mediated event. Autoimmunity 1993; 15:113-22.
38. Akhtar I, Gold JP, Pan L-Y et al. $CD4^+$ β islet cell-reactive T cell clones that suppress autoimmune diabetes in nonobese diabetic mice. J Exp Med 1995; 182:87-97.
39. van der Veen RC, Stohlman SA. Encephalitogenic Th1 cells are inhibited by Th2 cells with related peptide specificity: relative roles of interleukin (IL)-4 and IL-10. J Neuroimmunol 1993; 48:213-20.
40. Khoury SJ, Hancock WW, Weiner HL. Oral tolerance to myelin basic protein and natural recovery from experimental autoimmune encephalomyelitis are associated with downregulation of inflammatory cytokines and differential upregulation of transforming growth factor β, interleukin 4, and prostaglandin E expression in the brain. J Exp Med 1992; 176:1355-64.
41. Racke MK, Bonomo A, Scott DE et al. Cytokine-induced immune deviation as a therapy for inflammatory autoimmune disease. J Exp Med 1994; 180:1961-6.
42. Rapoport MJ, Jaramillo A, Zipris D et al. Interleukin 4 reverses T cell proliferative unresponsiveness and prevents the onset of diabetes in nonobese diabetic mice. J Exp Med 1993; 178:87-99.
43. Pennline KJ, Roquegaffney E, Monahan M. Recombinant human IL-10 prevents the onset of diabetes in the nonobese diabetic mouse. Clin Immunol Immunopathol 1994; 71:169-75.
44. Fowell D, Mason D. Evidence that the T cell repertoire of normal rats contains cells with the potential to cause diabetes. Characterization of the $CD4^+$ T cell subset that inhibits this autoimmune potential. J Exp Med 1993; 177:627-36.
45. Scott B, Liblau R, Degermann S et al. A role for non-MHC genetic polymorphism in susceptibility to spontaneous autoimmunity. Immunity 1994; 1:1-20.
46. Mueller R, Krahl T, Sarvetnick N. Pancreatic expression of interleukin-4 abrogates insulitis and autoimmune diabetes in nonobese diabetic (NOD) mice. J Exp Med 1996; 184:1093-9.
47. Healey D, Ozegbe P, Arden S et al. In vivo activity and in vitro specificity of CD4+ Th1 and Th2 cells derived from the spleens of diabetic NOD mice. J Clin Invest 1995; 95:2979-85.

48. Moritani M, Yoshimoto K, Tashiro F et al. Transgenic expression of IL-10 in pancreatic islet A cells accelerates autoimmune insulitis and diabetes in non-obese diabetic mice. Int Immunol 1994; 6:1927-36.
49. De Carli M, D'Elios M, Mariotti S et al. Cytolytic T cells with Th1-like cytokine profile predominate in retroorbital lymphocytic infiltrates of Graves' ophthalmopathy. J Clin Endocrinol Metab 1993; 77:1120-4.
50. Brod SA, Benjamin D, Hafler DA. Restricted T cell expression of IL-2, IFN-γ mRNA in human inflammatory disease. J Immunol 1991; 147:810-5.
51. Foulis AK, McGill M, Farquahrson MA. Insulitis in type I (insulin-dependent) diabetes mellitus in man. Macrophages, lymphocytes and interferon-γ-containing cells. J Pathol 1991; 165:97-103.
52. Miltenburg AM, van Laar JM, de Kuiper R et al. T cells cloned from human rheumatoid synovial membrane functionally represent the Th1 subset. Scand J Immunol 1992; 35:603-10.
53. De Carli M, D'Elios MM, Zancuoghi G et al. Human TH1 and TH2 cells: functional properties, regulation of development and role in autoimmunity. Autoimmunity 1994; 18:301-8.
54. Simon AK, Seipelt E, Sieper J. Divergent T-cell cytokine patterns in inflammatory arthritis. Proc Natl Acad Sci USA 1994; 91:8562-6.
55. Rocken M, Shevach EM. Immune deviation—the third dimension of nondeletional T cell tolerance. Immunol Rev 1996; 149:175-94.
56. Harrison LC. Islet cell autoantigens in insulin-dependent diabetes: Pandora's box re-visited. Immunol Today 1992; 13:348-52.
57. Kaufman DL, Clare-Salzler M, Tian J et al. Spontaneous loss of T-cell tolerance to glutamic acid decarboxylase in murine insulin-dependent diabetes. Nature 1993; 366:69-72.
58. Tisch R, Yang X-D, Singer SM et al. Immune response to glutamic acid decarboxylase correlates with insulitis in non-obese diabetic mice. Nature 1993; 366:72-5.
59. Tian J, Atkinson MA, Clare-Salzer M et al. Nasal administration of glutamate decarboxylase (GAD65) peptides induces Th2 responses and prevents murine insulin-dependent diabetes. J Exp Med 1996; 183:1561-7.
60. Acha-Orbea H, Mitchell DJ, Timmermann L et al. Limited heterogeneity of T cell receptors in experimental allergic encephalomyelitis. Cell 1988; 54:263-73.
61. Waisman A, Ruiz PJ, Hirschenberg DL et al. Suppressive vaccination with DNA encoding a variable region gene of the T-cell receptor prevents autoimmune encephalomyelitis and activates Th2 immunity. Nature Medicine 1996; 2:899-905.
62. Briner T, Kuo M-C, Keating KM et al. Peripheral T cell tolerance induced in naive and primed mice by subcutaneous injection of peptides from the major cat allergen Fel d I. Proc Natl Acad Sci USA 1993; 90:7608-12.

63. Wolf SF, Temple PA, Kobayashi M et al. Cloning of cDNA for natural killer stimulatory factor, a heterodimeric cytokine with multiple biological effects on T and natural killer cells. J Immunol 1991; 146:3074-81.
64. Gubler U, Chua AO, Schoenhaut DS et al. Coexpression of two distinct genes is required to generate secreted, bioactive cytotoxic lymphocyte maturation factor. Proc Natl Acad Sci USA 1991; 88:4143-7.
65. Trinchieri G. Interleukin-12: a cytokine produced by antigen-presenting cells with immunoregulatory functions in the generation of T-helper cells type 1 and cytotoxic lymphocytes. Blood 1994; 84:4008-27.
66. Germann T, Gately M, Schoenhaut DS et al. Interleukin-12/T cell stimulating factor, a cytokine with multiple effects on T helper type 1 (T_h1) but not on T_h2 cells. Eur J Immunol 1993; 23:1762-70.
67. Trembleau S, Penna G, Bosi E et al. IL-12 administration induces Th1 cells and accelerates autoimmune diabetes in NOD mice. J Exp Med 1995; 181:817-21.
68. O'Hara RM, Henderson SL. Interleukin 12 prevents the onset of diabetes in NOD mice. FASEB J 1995; 9:5938.
69. Magram J, Connaughton S, Warrier R et al. IL-12 deficient mice are defective in IFNγ production and type 1 cytokine responses. Immunity 1996; 4:471-82.
70. Leonard JP, Waldburger KE, Goldman SJ. Prevention of experimental autoimmune encephalomyelitis by antibodies against interleukin-12. J Exp Med 1995; 181:381-6.
71. Germann T, Szeliga J, Hess H et al. Administration of IL-12 in combination with type II collagen induces severe arthritis in DBA/1 mice. Proc Natl Acad Sci USA 1995; 92:4823-7.
72. Hess H, Gately MK, Ruede E et al. High doses of interleukin-12 inhibit the development of joint disease in DBA/1 mice immunized with type II collagen in complete Freund's adjuvant. Eur J Immunol 1996; 26:187-91.
73. Neurath MF, Fuss I, Kelsall BL et al. Antibodies to interleukin 12 abrogate established experimental colitis in mice. J Exp Med 1995; 182:1281-90.
74. Chua AO, Chizzonite R, Desai BB et al. Expression cloning of a human IL-12 receptor component. A new member of the cytokine receptor superfamily with strong homology to gp130. J Immunol 1994; 153:128-36.
75. Chua AO, Wilkinson VL, Presky DH et al. Cloning and characterization of a mouse IL-12 receptor β component. J Immunol 1995; 155:4286-94.
76. Murphy EE, Terres G, Macatonia SE et al. B7 and interleukin 12 cooperate for proliferation and interferon γ production by mouse T helper clones that are unresponsive to B7 costimulation. J Exp Med 1994; 180:223-31.
77. Ding L, Linsley PS, Huang LY et al. IL-10 inhibits macrophage costimulatory activity by selectively inhibiting the up-regulation of B7 expression. J Immunol 1993; 151:1224-34.
78. Fiorentino DF, Zlotnik A, Mosmann TR et al. IL-10 inhibits cytokine production by activated macrophages. J Immunol 1991; 147:3815-622.

79. de Waal Malefyt R, Abrams J, Bennett B et al. IL-10 inhibits cytokine synthesis by human monocytes: an auto-regulatory role of IL-10 produced by monocytes. J Exp Med 1991; 174:1209-20.
80. Bogdan C, Vodovotz Y, Nathan C. Macrophage deactivation by interleukin 10. J Exp Med 1991; 174:1549-55.
81. Groux H, Bigler M, De Vries JE et al. Interleukin-10 induces a long-term antigen specific anergic state in human $CD4^+$ T cells. J Exp Med 1996; 184:19-29.
82. Bromberg JS. IL-10 immunosuppression in transplantation. Curr Opin Immunol 1995; 7:639-43.
83. Kuehn R, Loehler J, Rennick D et al. Interleukin-10-deficient mice develop chronic enterocolitis. Cell 1993; 75:263-74.
84. Powrie F, Leach MW, Mauze S et al. Inhibition of Thl responses prevents inflammatory bowel disease in *scid* mice reconstituted with $CD45RB^{hi}$ $CD4^+$ T cells. Immunity 1994; 1:553-62.
85. Chatelain R, Varkila K, Coffman RL. IL-4 induces a Th2 response in *Leishmania major*-infected mice. J Immunol 1992; 148:1182-7.
86. Sadick MD, Heinzel FP, Holaday BJ et al. Cure of murine leishmaniasis with anti-interleukin-4 monoclonal antibody. Evidence for a T cell-dependent mechanism. J Exp Med 1990; 171:115-27.
87. Powrie F, Menon S, Coffman RL. Interleukin-4 and interleukin-10 synergize to inhibit cell-mediated immunity in vivo. Eur J Immunol 1993; 23:3043-9.
88. Shehadeh N, Calcinaro F, Bradley BJ et al. Effect of adjuvant therapy on development of diabetes in mouse and man. Lancet 1994; 343:706-7.
89. Rabinovitch A, Suarez-Pinzon WL, Sorensen O et al. IFN-γ gene expression in pancreatic islet-infiltrating mononuclear cells correlates with autoimmune diabetes in nonobese diabetic mice. J Immunol 1995; 154:4874-82.
90. Allen JB, Wong HL, Costa GL et al. Suppression of monocyte function and differential regulation of IL-1 and IL-1ra by IL-4 contribute to resolution of experimental arthritis. J Immunol 1993; 151:4344-51.
91. Constant S, Pfeiffer C, Woodard A et al. Extent of T cell receptor ligation can determine the functional differentiation of naive CD4+ T cells. J Exp Med 1995; 182:1591-6.
92. Hosken NA, Shibuya K, Heath AW et al. The effect of antigen dose on $CD4^+$ helper cell phenotype development in a T cell receptor-αβ-transgenic model. J Exp Med 1995; 182:1579-84.
93. Guéry J-C, Galbiati F, Smiroldo S et al. Selective development of Th2 cells induced by continuous administration of low dose soluble proteins to normal and β2-microglobulin-deficient BALB/c mice. J Exp Med 1996; 183:485-97.
94. Weiner HL, Friedman A, Miller A et al. Oral tolerance: immunologic mechanisms and treatment of animal and human organ-specific autoimmune diseases by oral administration of autoantigens. Annu Rev Immunol 1994; 12:809-37.

95. Burstein HJ, Shea CM, Abbas AK. Aqueous antigens induce in vivo tolerance selectively in IL-2- and IFN-γ-producing (Th1) cells. J Immunol 1992; 148:3687-91.
96. De Wit D, Van Mechelen M, Ryelandt M et al. The injection of deaggregated gamma globulins in adult mice induces antigen-specific unresponsiveness of T helper type 1 but not type 2 lymphocytes. J Exp Med 1992; 175:9-14.
97. Yoshimoto T, Paul EP. $CD4^{pos}$, $NK1.1^{pos}$ T cells promptly produce interleukin 4 in response to in vivo challenge with anti-CD3. J Exp Med 1994; 179:1285-95.
98. Bendelac A, Killeen N, Littman DL et al. A subset of $CD4^+$ thymocytes selected by MHC clas I molecules. Science 1994; 263:1774-8.
99. Coles MC, Raulet DH. Class I dependence of development of $CD4^+$ $CD8^-$ $NK1.1^+$ thymocytes. J Exp Med 1994; 180:395-9.
100. Launois P, Okteki T, Swihart K et al. In susceptible mice, *Leishmania major* induce very rapid interleukin-4 production by $CD4^+$ T cells which are $NK1.1^-$. Eur J Immunol 1995; 25:3298-307.
101. Stockinger B, Zal T, Zal A et al. B cells solicit their own help from T cells. J Exp Med 1996; 183:891-9.
102. Day MJ, Tse AGD, Puklavec M et al. Targeting autoantigen to B cells prevents the induction of a cell-mediated autoimmune disease in rats. J Exp Med 1992; 175:655-9.
103. Saoudi A, Simmonds S, Huitinga I et al. Prevention of experimental allergic encephalomyelitis in rats by targeting autoantigen to B cells: evidence that the protective mechanism depends on changes in the cytokine response and migratory properties of autoantigen-specific T cells. J Exp Med 1995; 182:335-44.
104. Eynon EE, Parker DC. Parameters of tolerance induction by antigen targeted to B lymphocytes. J Immunol 1993; 151:2958-64.
105. Degermann S, Pria E, Adorini L. Soluble protein but not peptide administration diverts the immune response of a clonal T cell population to the Th2 pathway. J Immunol 1996;
106. Crowley MT, Inaba K, Steinman RM. Dendritic cells are the principal cells in mouse spleen bearing immunogenic fragments of foreign proteins. J Exp Med 1990; 172:383-6.
107. Constant S, Sant'Angelo D, Pasqualini T et al. Peptide and protein antigen require distinct antigen-presenting cell subsets for the priming of $CD4^+$ T cells. J Immunol 1995; 154:4915-23.
108. Constant S, Schweitzer N, West J et al. B lymphocytes can be competent antigen-presenting cells for priming $CD4^+$ T cells to protein antigens in vivo. J Immunol 1995; 155:3734-41.
109. Guéry J-C, Ria F, Galbiati F et al. The mode of protein antigen administration determines preferential presentation of peptide-class II complexes by lymph node dendritic or B cells. Int Immunol 1997; 9:9-15.

110. Nelson CA, Petzold SJ, Unanue ER. Peptides determine the lifespan of MHC class II molecules in the antigen-presenting cell. Nature 1994; 371:250-2.
111. Nicholson LB, Greer JM, Sobel RA et al. An altered peptide ligand mediates immune deviation and prevents autoimmune encephalomyelitis. Immunity 1995; 3:397-405.
112. Pfeiffer C, Stein J, Southwood S et al. Altered peptide ligands can control CD4 T lymphocyte differentiation in vivo. J Exp Med 1995; 181:1569-74.
113. Windhagen A, Scholz C, Hollsberg P et al. Modulation of cytokine patterns of human autoreactive T cell clones by a single amino acid substitution of their peptide ligand. Immunity 1995; 2:373-80.
114. Lider O, Santos LMB, Lee CSY et al. Suppression of experimental autoimmune encephalomyelitis by oral administration of myelin basic protein II. Suppression of disease and in vitro response is mediated by CD8+ T lymphocytes. J Immunol 1989; 142:748-754.
115. Miller A, Lider O, Roberts AB et al. Suppressor T cells generated by oral tolerization to myelin basic protein suppress both in vitro and in vivo immune responses by the release of transforming growth factor beta after antigen-specific triggering. Proc Natl Acad Sci USA 1992; 89:421-5.
116. O'Garra A, Murphy K. Role of cytokines in determining T-lymphocyte function. Curr Opin Immunol 1994; 6:458-66.
117. Chen Y, Kuchroo VK, Inobe J-I et al. Regulatory T cell clones induced by oral tolerance: suppression of autoimmune encephalomyelitis. Science 1994; 265:1237-40.
118. Lehmann P, Sercarz E, Forsthuber T et al. Determinant spreading and the dynamics of the autoimmune T cell repertoire. Immunol Today 1993; 14:203-8.
119. Chen Y, Inobe J-I, Marks R et al. Peripheral deletion of antigen-reactive T cells in oral tolerance. Nature 1995; 376:177-80.
120. Critchfield JM, Racke MK, Zuniga-Pflucker JC et al. T cell deletion in high antigen dose therapy of autoimmune encephalomyelitis. Science 1994; 263:1139-43.
121. Weinberg AD, Whitham R, Swain SL et al. Transforming growth factor beta enhances the in vivo effector function and memory phenotype of antigen-specific T helper cells in experimental autoimmune encephalomyelitis. J Immunol 1992; 148:2109-17.
122. Metzler B, Wraith DC. Inhibition of experimental autoimmune encephalitis by inhalation but not by oral administration of encephalitogenic peptide: influence of MHC binding. Int Immunol 1992; 5:1159-65.
123. Weiner HL, Mackin GA, Matsui M et al. Double-blind pilot trial of oral tolerization with myelin antigens in multiple sclerosis. Science 1993; 259:1321-4.
124. Trentham DE, Dynesius-Trentham RA, Orav EJ et al. Effects of oral administration of type II collagen on rheumatoid arthritis. Science 1993;

261:1727-30.
125. Evavold BD, Allen PM. Separation of IL-4 production from Th cell proliferation by an altered T cell receptor ligand. Science 1991; 252:1308-10.
126. Seder RA, Paul WE, Davis MM et al. The presence of interleukin 4 during in vitro priming determines the lymphokine-producing potential of CD4+ T cells from T cell receptor transgenic mice. J Exp Med 1992; 176:1091-8.
127. Hsieh CS, Heimberger A, Gold J et al. Differential regulation of T helper phenotype development by IL-4 and IL-10 in α, β T-cell receptor transgenic system. Proc Natl Acad Sci USA 1992; 89:6065-9.
128. Evavold BD, Sloan-Lancaster J, Allen PM. Tickling the TCR: selective T cell functions stimulated by altered peptide ligands. Immunol Today 1993; 14:602-9.
129. Wegener AM, Letourner F, Hoeveler A et al. The T cell receptor/CD3 complex is composed of at least two autonomous transduction modules. Cell 1992; 68:83-95.
130. Sloan-Lancaster J, Shaw AS, Rothbard JB et al. Partial T cell signaling: altered phospho-ζ and lack of zap70 recruitment in APL-induced T cell anergy. Cell 1994; 79:913-22.
131. Madrenas J, Wange RL, Wang JL et al. ζ phosphorylation without zap-70 activation induced by TCR antagonists or partial agonists. Science 1995; 267:515-8.
132. Sloan-Lancaster J, Evavold BD, Allen PM. Induction of T cell anergy by altered T cell receptor ligand on live antigen-presenting cells. Nature 1993; 363:156-9.
133. Sloan-Lancaster J, Evavold B, Allen PM. Th2 cell clonal anergy as a consequence of partial activation. J Exp Med 1994; 180:1195-205.
134. De Magistris TM, Alexander J, Coggeshall M et al. Antigen analog-major histocompatibility complexes act as antagonists of the T cell receptor. Cell 1992; 68:625-34.
135. Jameson SC, Carbone FR, Bevan MJ. Clone-specific T cell receptor antagonists of major histocompatibility complex class I-restricted cytotoxic T cells. J Exp Med 1993; 177:1541-50.
136. Klenerman P, Rowland-Jones S, McAdam S et al. Cytotoxic T-cell activity antagonized by naturally occurring HIV-1 Gag variants. Nature 1994; 369:403-7.
137. Bertoletti A, Sette A, Chisari FV et al. Natural variants of cytotoxic epitopes are T-cell receptor antagonists for antiviral cytotoxic T cells. Nature 1994; 369:407-10.
138. Franco A, Southwood S, Arrhenius T et al. T cell receptor antagonist peptides are effective inhibitors of experimental allergic encephalomyelitis. Eur J Immunol 1994; 24:940-6.
139. Karin N, Mitchell DJ, Brocke S et al. Reversal of experimental autoimmune encephalomyelitis by a soluble peptide variant of a myelin basic protein epitope: T cell receptor antagonism and reduction of interferon

γ and tumor necrosis factor α production. J Exp Med 1994; 180:2227-37.

140. Brocke S, Gijbels K, Allegretta M et al. Treatment of experimental encephalomyelitis with a peptide analogue of myelin basic protein. Nature 1996; 379:343-6.
141. Smilek DE, Wraith DC, Hodgkinson S et al. A single amino acid change in myelin basic protein peptide confers the capacity to prevent rather than induce experimental autoimmune encephalomyelitis. Proc Natl Acad Sci USA 1991; 88:9633-7.
142. Gautam AM, Pearson CI, Smilek DE et al. A polyalanine peptide with only five native myelin basic protein residues induces autoimmune encephalomyelitis. J Exp Med 1992; 176:605-9.
143. Wauben MHM, Boog CJP, van der Zee R et al. Disease inhibition by major histocompatibility complex binding peptide analogues of disease-associated epitopes: more than blocking alone. J Exp Med 1992; 176:667-77.
144. Kingsley G, Lanchbury J, Panayi G. Immunotherapy in rheumatic disease: an idea whose time has come—or gone? Immunol Today 1996; 17:9-12.
145. Keffer J, Probert L, Cazlaris H et al. Transgenic mice expressing human tumor necrosis factor—a predictive genetic model of arthritis. EMBO J 1991; 13:4025-31.
146. Williams RO, Feldmann M, Maini RN. Anti-tumor necrosis factor ameliorates joint disease in murine collagen-induced arthritis. Proc Natl Acad Sci USA 1992; 89:9784-8.
147. Elliott MJ, Maini RN, Feldmann M et al. Treatment of rheumatoid arthritis with chimeric monoclonal antibodies to tumor necrosis factor α. Arthritis Rheum 1993; 36:1681-90.
148. Haak-Frendscho M, Marsters SA, Mordenti J et al. Inhibition of TNF by a TNF receptor immunoadhesin. Comparison to an anti-TNF monoclonal antibody. J Immunol 1994; 152:1347-53.
149. Baker D, Butler D, Scallon BJ et al. Control of established experimental allergic encephalomyelitis by inhibition of tumor necrosis factor (TNF) activity within the central nervous system using monoclonal antibodies and TNF receptor-immunoglobulin fusion proteins. Eur J Immunol 1994; 24:2040-8.
150. Heinzel FP, Schoenhaut DS, Rerko RM et al. Recombinant interleukin 12 cures mice infected with *Leishmania major*. J Exp Med 1993; 177:1505-9.